Hormones, Cells
and Organisms

Hormones, Cells and Organisms
The role of hormones in mammals

P. Catherine Clegg
B.Sc., Ph.D., M.B., Ch.B.

Sometime Lecturer in Physiology
University of Sheffield

and

Arthur G. Clegg
B.Sc., M.I.Biol.

Sometime Principal Lecturer in Biology
City College of Education, Sheffield

Heinemann Educational Books Ltd
London

Heinemann Educational Books Ltd
LONDON MELBOURNE TORONTO AUCKLAND
SINGAPORE HONG KONG JOHANNESBURG
IBADAN NAIROBI

SBN 435 62170
© P. C. and A. G. Clegg 1969
First published 1969

Published by Heinemann Educational Books Ltd
48 Charles Street, London W.1. X 8AH
Printed in Great Britain by
Richard Clay (The Chaucer Press) Ltd
Bungay, Suffolk

Preface

We have written this book in the hope that it will serve as a useful introduction to the student who is beginning a study of endocrinology. No attempt has been made to give a comprehensive cover of all endocrine tissues and their hormones. Instead we have made a representative selection of material from the vast and rapidly expanding literature of endocrinology to illustrate various themes such as the rôle of hormones in homeostasis and adaptation to the environment. Wherever practicable an attempt has been made to illustrate the experimental basis of the subject and in addition there is a separate chapter on techniques in endocrine research. Recent advances in molecular biology and microbial genetics have stimulated rapid advances in the study of the mechanisms of hormone action at the molecular level and no apology is given for the apparently disproportionate length of our account of this new and fascinating aspect of endocrinology. In a book of this size it has been impossible to deal with comparative aspects of the subject. The only non-mammalian hormone we have discussed is the insect hormone ecdysone, and this is included only because it illustrates in such a dramatic fashion the action of a hormone on the genetic material of the cell.

BRIGHTON, 1969.
A. G. C.
P. C. C.

Acknowledgements

We are very grateful to the following who have given us permission to reproduce their photographs or illustrations:

Mr W. H. Freeman and Mr B. Bracegirdle (Plates 1, 2, 3a, 3b, 4a, 4b), Professor A. A. Harper (fig. 37), Professor I. MacIntyre (fig. 18), Professor P. D. Ray (fig. 13).

Contents

List of Plates

1

The Nature of Hormones

Introduction: The meaning of the word hormone

In the early 17th century William Harvey published his studies which demonstrated, for the first time, the fact of the circulation of the blood.[1] He showed that the heart is a double-chambered pump. From one chamber (the left ventricle) blood is delivered to the systemic arteries which supply blood to all the tissues of the body except the lungs. Blood is returned to the second chamber of the heart (the right ventricle) from the tissues, by way of a second system of vessels, the veins. Blood is returned to the first chamber through the pulmonary circulation. Along this system of chambers and vessels there is a constant flow of blood. Until Harvey's time it had, for example, been thought that the arteries contained air, an error which he pointed out was due to the fact that animals for dissection had been killed by bleeding them to death, thus emptying the arteries. In addition to the dead carcases, Harvey studied the living intact animal and he noted, for example, that when the arteries were cut blood gushed from them under pressure.

This discovery of the circulation of the blood was critical to the development of physiological thought and without it a proper understanding of the processes of respiration, digestion, absorption and excretion would have been impossible. However, it was over two hundred years later that biologists discovered that in this circulating blood chemical substances were transported from particular tissues or organs producing them, so as to exert profound effects on distant tissues. One of the earliest hints of this function of the circulating blood was provided by the observations of Berthold. The effects of removing the testes from cockerels had been known for centuries. When

[1] *The Motion of the Heart and Blood in Animals*, published in Latin 1628.

this operation is done the comb and wattles of the cock shrink in size and lose their turgidity and brilliant colouring (fig. 1). The characteristic male plumage disappears, aggressive behaviour is lost, and the animal becomes plump and suitable for table meat. In 1849 Berthold described experiments in which he had transplanted testes from normal cocks into capons (castrated cocks). The capons then redeveloped, to a degree, the characteristic features of the cock bird. These results suggested that some active substance was liberated from the transplanted testes and this substance reached the organs of the

FIGURE 1. The effect of castration on the comb and wattles of the cockerel. Normal cockerel (*left*) and the effect of castration.

capon by way of the circulating blood. It was later found that injection of testicular extracts could produce the same effects as transplantation of the testes. In 1927 McGee isolated a relatively pure chemical substance from the testes of bulls which could produce similar masculinizing, or androgenic, effects when injected into animals. Pure crystalline androgens were later synthesized. These substances are highly potent and produce effects when given to animals in doses of micro-grams. When minute quantities, for example, are rubbed into the base of the bill of a sexually immature male sparrow the bird soon develops the black bill typical of the sexually mature male.

In the late 19th and early 20th centuries a great variety of organ and tissue extracts were found to be capable of producing marked physiological effects when injected into the blood stream of animals. They have been found in a wide range of species of both invertebrate (insects, custaceans, molluscs) and vertebrate (fishes, amphibia, reptiles, birds and mammals) types. The active chemical substances present in these extracts have been given the name of *hormones*. The word hormone (meaning 'I arouse') was first used by Starling in 1905. He found that when an extract of duodenal mucosa was injected into the vein of an animal there followed a flow of alkaline secretion from the pancreas. The hormone, the active substance present in the duodenal extract, was given the name of secretin. Later there followed the discovery of a variety of different hormones produced by different parts of the bowel wall and intimately concerned in the regulation of secretory and motor activity of the bowel and adjacent secretory glands (chapter 13). Today there are literally hundreds of chemical substances which have been isolated from various tissues and organs which can produce distinct effects on growth, secretion, metabolism or behaviour when they are injected into the blood stream. Many hormones will produce their effects when applied to organs isolated from the body or even when added to suspensions of tissue homogenate or to subcellular preparations such as isolated mitochondria. To demonstrate that these substances are involved in the normal regulation of function in the intact animal, many complex experiments may be required and these will be considered in chapter 2.

Local and general hormones

In 1935 Huxley made the first attempt at the classification of the bewildering array of active substances isolated from animal tissues. He divided these active chemical substances, or activators, into two broad groups.

1. Local activators.

2. Distance activators $\begin{cases} \text{diffusion activators.} \\ \text{circulating activators.} \end{cases}$

A vast number of chemical substances are synthesized within living cells. We now know that these synthetic processes are ultimately regulated by the gene material (deoxyribose nucleic acid) of the chromosomes. Many of these substances, e.g. the enzymes which control the rate of chemical reactions, are restricted to the cells producing them. These are included in the *local activators*. In some cells some of the substances synthesized inside the cell ultimately leave the cell (i.e. they are secreted) and have effects on other cells and tissues. These substances may reach their 'target cells' by diffusion and their sphere of action may be relatively restricted. Huxley termed these *diffusion activators*. In this class one may include substances such as kallidins, histamine, acetyl choline, noradrenaline, gamma-aminobutyric acid, and those, as yet chemically unidentified, substances called organizers which are responsible for regulating the patterns of organogenesis in the early embryo. Another group of activators, called *circulating activators*, are distributed throughout the body in the circulating blood. It is these circulating activators which have come to be regarded as 'hormones', but there is no clear line of demarcation between diffusion and circulating activators. Indeed, some substances may belong to both groups. Sympathetic nerves exert their effects on the structures they innervate by the release of small amounts of a highly active substance, noradrenaline, from the nerve terminals. The noradrenaline diffuses from the terminals across the small distance which separates the nerve and the end organ and exerts its effects upon the cells. These nerves mediate their effects by the liberation of a chemical substance and indeed the application of noradrenaline to a tissue produces effects which mimic those of sympathetic nerve stimulation. Noradrenaline is also liberated in larger amounts from the medulla, or core, of the adrenal gland and is distributed by the blood stream throughout the body where it has widespread effects, particularly upon the heart and blood vessels. It may thus be best not to restrict the term hormone to substances transported in the circulating blood but to divide hormones into two general groups, *local hormones* and *general hormones*. Examples of local hormones have already been given. Many hormones fall into the class of general hormones, including adrenaline, noradrenaline, thyroxine, oestrogens, androgens, cortisol, aldosterone, insulin, and the hormones secreted by the pituitary gland.

Sources of hormones

Glands

Hormones have been isolated from many sites in the vertebrate organism. Perhaps the most familiar sites are those highly organized structures which function as secretory units that are called glands. Examples of these structures include the pituitary, adrenal, thyroid and parathyroid glands. Each has its own blood and lymphatic supply. Because the secretions of these glands are discharged into the blood stream and lymph, these glands have no ducts, hence the names *ductless* or *endocrine* glands, contrasting with such structures as the salivary glands which discharge their secretions into the mouth by way of a duct. These latter glands, with which one may include the pancreas, liver, sebaceous and sweat glands, are termed *exocrine glands* because they discharge their secretions to the 'exterior' of the body. Endocrine glands characteristically have a rich blood supply, the blood serving not only as a vehicle of transport for the secretions but also as a source of the chemical raw materials from which the hormones are synthesized. The thyroid gland has, weight for weight, a richer blood supply than the kidney, which takes about a quarter of the entire output of the heart each minute.

Non-glandular sources of hormones

Hormones have been isolated, not only from typically organized endocrine glands but also from a great variety of tissues and cells scattered throughout the body. For example, although the hormones adrenaline and noradrenaline are produced by the adrenal medulla, which is an organized glandular structure, they are also produced by groups of medullary type cells which are scattered through many tissues. These medullary type cells are called chromaffine cells, and are particularly associated with the larger blood vessels of the abdomen and the pelvis. The hormones oxytocin and vasopressin are produced by groups of neurosecretory cells lying in the hypothalamus in the floor of the thalamencephalon (fig. 5). The hormones insulin and glucagon are synthesized within small groups of cells called islets of Langerhans which are intermingled with the exocrine secreting tissue of the pancreas (plate 1). Between the seminiferous tubules of the testis are the interstitial cells which produce the male sex hormones (plate 3b). The hormone histamine is distributed widely in various tissues such as kidney, liver, pancreas, stomach and parts of the peripheral nervous system. Much of the histamine present in these tissues is found in mast cells, which contain the enzyme histidine decarboxylase, thus enabling them to convert the amino-acid histidine into histamine. The raw material for the production of the local hormone bradykinin is distributed throughout all the tissues of the body in the form of a precursor, bradykininogen, which is converted to bradykinin by locally liberated proteolytic enzymes.

Although the precise cellular origin of very many hormones is known, there are other hormones about which we have, at present, no precise information as to the location of synthesis and release. This is particularly true of a more recently discovered group of hormones, the lipid anions, which are present in many tissues such as the central nervous system, iris, lung, uterus and seminal vesicle. Even in the case of hormones which have been known for a long time we do not always know the precise cellular origin. For example, we are not sure exactly which cells in the ovary produce the female hormone oestrogen.

The chemical nature of hormones

The many hormones which have been chemically identified fall into one of several classes of substances including proteins and polypeptides, amino-acid derivatives, lipids and sterols.

Proteins and polypeptides

These substances consist of amino acids linked by peptide bonds into long-chain molecules. The number of constituent amino acids and thus the molecular weight varies greatly from one hormone to another. Those containing fewer than a hundred amino acids are classified as polypeptides. The smallest polypeptide hormones are oxytocin and vasopressin (eight amino acids), kallidin I or bradykinin (nine amino acids), kallidin II (ten amino acids) and insulin (fifty-one

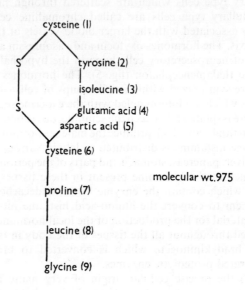

FIGURE 2. The chemical structure of oxytocin (bovine or pig).

amino acids). Vasopressin, found in monotremes, marsupials and most placental mammals, has a very similar structure to that of oxytocin which is shown in figure 2, except that phenylalanine replaces isoleucine in position (3) and arginine replaces leucine in position (8). Although the structure of the two hormones is so very similar they have very different effects, oxytocin regulating the contraction of uterine muscle and vasopressin acting on the kidney tubules regulating the conservation of water.

Some hormones are complex proteins e.g. those secreted by the anterior pituitary gland, including prolactin, follicle stimulating

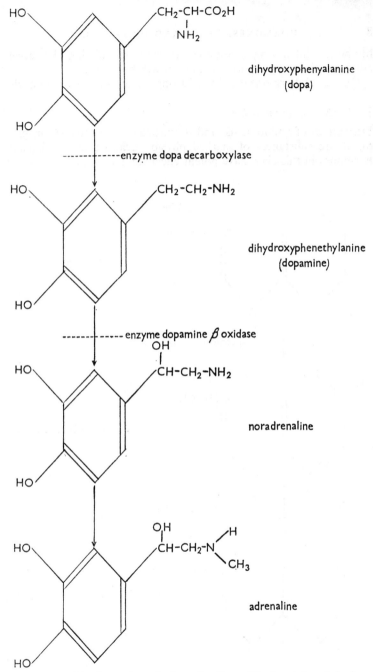

FIGURE 3. The formation of adrenaline from dihydroxyphenylalanine (DOPA)

hormone, luteinizing hormone, thyroid stimulating hormone, adrenocorticotrophic hormone and growth hormone. Ox-growth hormone consists of a single chain of 396 amino acids.

Derivatives of amino acids

Derivatives of amino acids, and in particular decarboxylation products, are substances of marked physiological activity, and many hormones in this class have potent effects on smooth muscle.

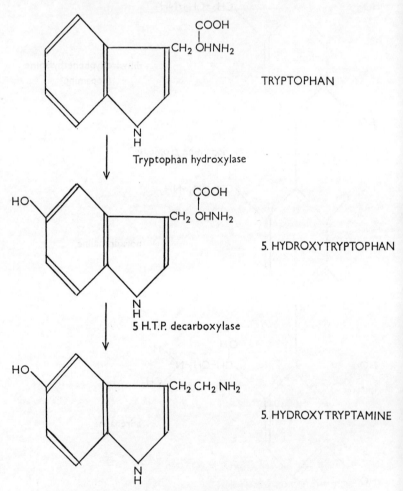

FIGURE 4. The formation of 5-hydroxytryptamine from tryptophan

NORADRENALINE AND ADRENALINE
The hormones noradrenaline and adrenaline are derived from the amino acids tyrosine and dihydroxyphenylalanine (DOPA). Decarboxylation of the amino acid DOPA produces the amine dopamine, which is then converted into the active hormone noradrenaline. Methylation of noradrenaline produces adrenalin (fig. 3).

5-HYDROXYTRYPTAMINE
5-Hydroxytryptamine (5HT) is derived initially from the amino acid tryptophan which is first hydrolysed to 5-hydroxytryptophan (5HTP). Decarboxylation of 5HTP results in the formation of 5HT (fig. 4).

HISTAMINE
Another local hormone having potent effects on smooth muscle is histamine, which is formed, particularly in the mast cells of tissues, by decarboxylation of the amino-acid histidine.

$$HC \underset{N-C \cdot CH_2CH_2NH_2}{\overset{NH-CX}{\big<}}$$

Histamine

γ-AMINO BUTYRIC ACID (GABA)
This substance has, for a long time, been known to be a constituent of green plants and fungi. In 1950 it was first isolated from the mammalian brain. It is produced by the decarboxylation of glutamic acid.

$$H_2NCHCOOH \qquad\qquad H_2NCH_2$$
$$CH_2 \quad \xrightarrow{-CO_2} \quad CH_2$$
$$CH_2COOH \qquad\qquad CH_2COOH$$
Glutamic acid   GABA

In crayfish muscle inhibitor nerve fibres may exert their effects by the release of GABA at the nerve terminals, and there is the probability that some inhibitory nerve terminals within the central nervous system of mammals may exert their effects on other neurones by means of this substance.

Lipids (esters of higher aliphatic acids)
One group of lipids which has been isolated from animal tissues has potent smooth muscle stimulating activity and may well be hormonal

in nature. Such active lipids have, for example, been isolated from brain, iris, intestine, lung, endometrium, menstrual fluid and semen.

One group of these active lipids is called 'prostaglandins' because they were first found in semen and were thought to arise in the prostate gland. The structure of one prostaglandin from the sheep vesicular gland is as follows:

$$CH_3(CH_2)_4 \ CHOH-CH{=}CH-CH-CH(CH_2)_6-COOH$$
$$HOCH \quad C-O$$
$$CH_2$$

Substances of a similar nature (nitrogen-free unsaturated hydroxy-fatty acids) have been isolated from a variety of other tissues. Human menstrual fluid contains at least two prostaglandins and it has been suggested that these substances are liberated into the blood stream from the endometrium during menstruation and are transported to the muscular part of the uterus where they induce rhythmical contractions which assist in the expulsion of the menstrual fluid.

If one removes a piece of frog intestine and suspends it in a bath of physiological solution, then a substance appears in the bathing solution which will stimulate other smooth muscles. This substance is a lipid. If the vagus nerve is stimulated, then there is an increased loss of this lipid (called darmstoff) and its function may be that of a local hormone, acting as a neuromuscular transmitting substance.

However, the full physiological significance of this exciting group of active lipids remains to be resolved. They do provide a rather interesting example of a chemical substance produced by one individual and capable of producing effects in another individual of the same species. Human semen is a rich source of prostaglandins and after coitus they may be absorbed from the female genital tract and produce changes in the activity of the uterus and Fallopian tubes. These changes in activity may be important in determining the degree of fertility of the female. The semen of many infertile or subfertile men is low in prostaglandin content.

STEROLS

These are a very important group of lipid substances which are extracted from animal tissues using fat solvents. Unlike many lipids they are non-saponifiable. They are crystalline cyclic alcohols occurring as the free compound or esterified with long-chain fatty acids. They all contain a saturated phenanthrene ring system to which is fused an additional five-membered ring.

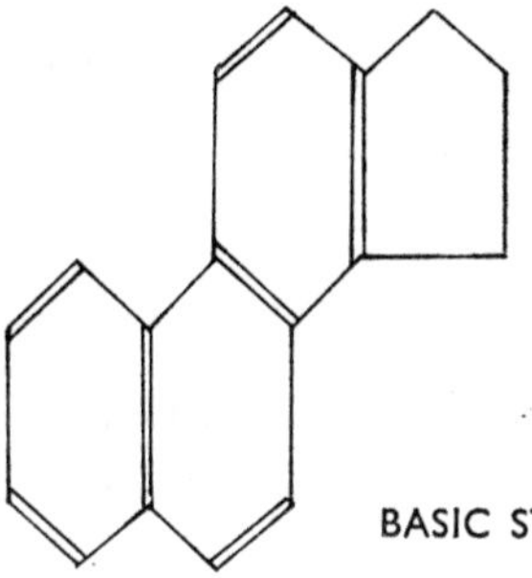

BASIC STEROL STRUCTURE

Cholesterol was the first of this group of compounds to be recognized:

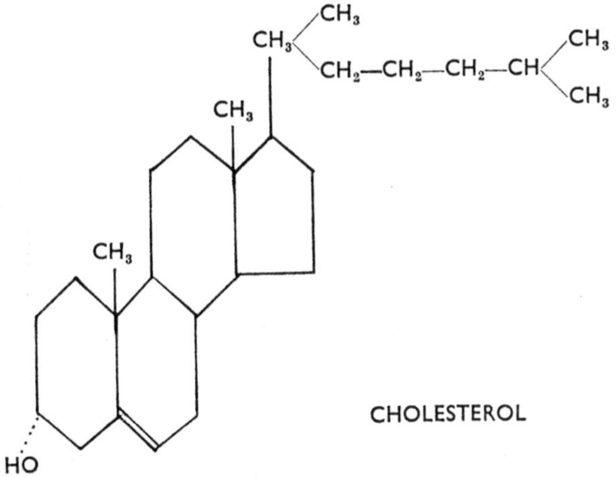

CHOLESTEROL

Cholesterol occurs free and esterified, in the blood and all animal tissues. It forms the raw material for the synthesis of many sterol hormones, including male and female sex hormones and the hormones of the adrenal cortex. The structural formulae for some of the more important sterol hormones is shown below.

OESTRADIOL

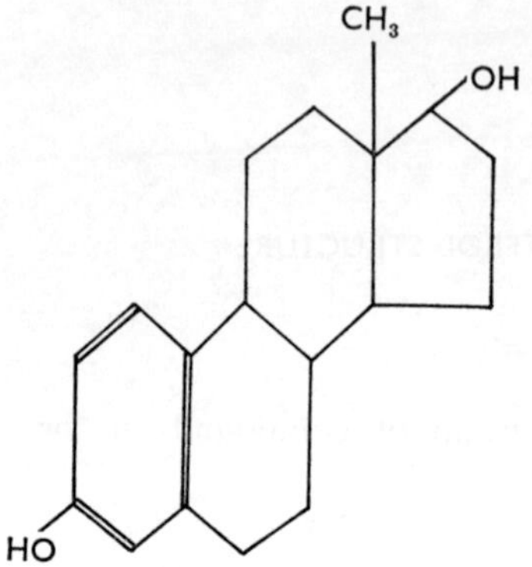

HYDROCORTISONE

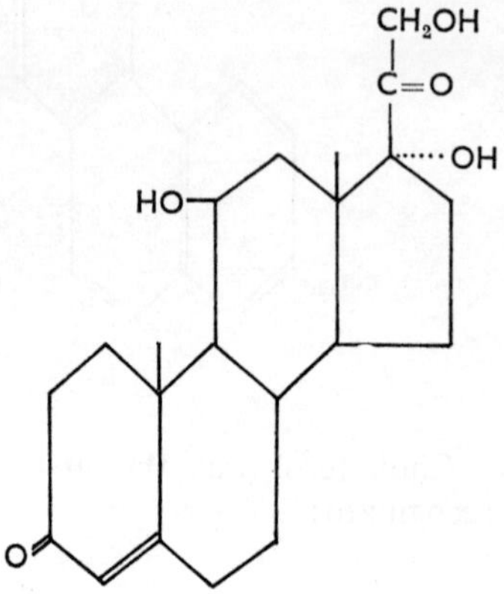

ALDOSTERONE

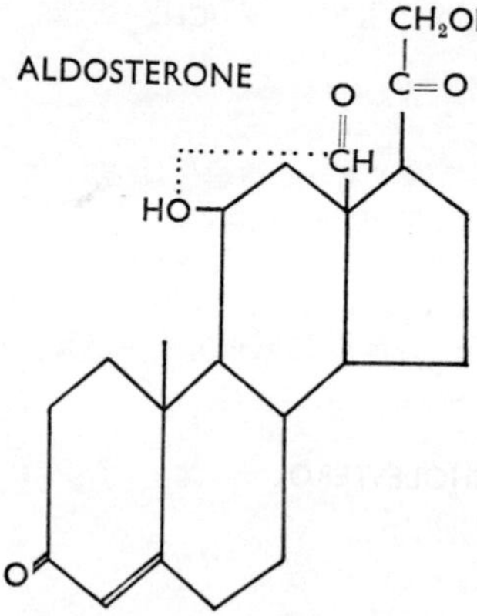

TESTOSTERONE

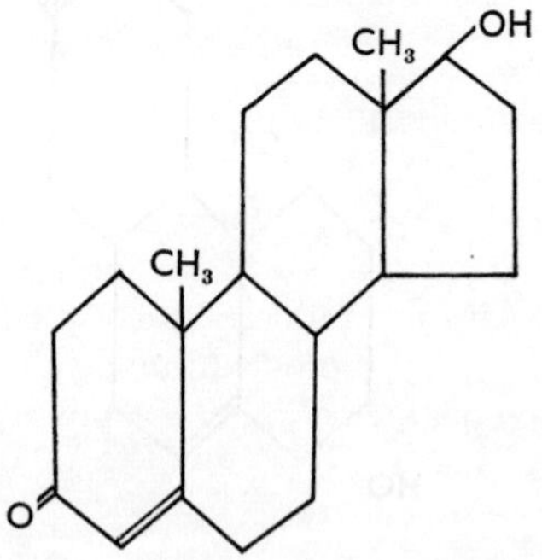

2

Techniques in Endocrine Research

The following may be considered to be an ideal sequence of studies in the exploration of an endocrine gland.

1. A morphological and histological identification of the tissue or organ.
2. A study of the effect of removal of the organ.
3. A study of the effects of replacing the removed organ by transplantation or by the injection of organ extracts, or by injection of the purified hormone.
4. The isolation of active principle(s) from the organ extract and the determination of its chemical structure.

Possibilities now arise for synthesis of the hormone and for the provision of relatively large amounts of pure hormone which may be used in more refined experimental situations in order to study the characteristics and possible mode of action of the hormone.

This is an ideal sequence of exploration and a study of the history of endocrinological research shows that it has not always been practicable, nor indeed possible, to follow this logical sequence of investigation.

Morphological studies

An outstanding anatomical feature of an endocrine gland is the absence of a duct in a structure with an obviously glandular appearance. The histological appearance confirms the glandular nature of the organ, with a fundamental structure consisting of groups of cells surrounded by meshes of blood capillaries. Typical endocrine cells show a rich and diverse cytoplasmic structure related to their secretory function. Using classical microscopic techniques one can demonstrate the presence of granules, mitochondria, Golgi apparatus and a rich ergoplaston, features all of which reflect the intense synthetic activities proceeding in the endocrine cell. Some of these cellular inclusions demonstrable by classical techniques have in the past been regarded as artefacts. However, more modern techniques using phase contrast and electron microscopy have confirmed these inclusions as real components of the secretory cell.

More recent techniques, including those of autoradiography and antibody staining technique have permitted a more accurate localization of hormone production, so that one can locate precisely which cells in an organ are actually producing a particular hormone. An example of a recent elegant technique used in hormone localization is that of fluorescent antibody staining.

Fluorescent antibody staining

When a foreign material is injected into an animal, the animal may respond by producing an antibody against the foreign material. The foreign material is called an antigen, and most antigens have a molecular weight greater than 5,000. The antibody produced by the animal in response to the introduction of an antigen is a protein, appearing in the globulin fraction of the blood proteins. The antibody has a specific affinity for the antigen which excited its production. One can prepare antibodies against many hormones, particularly those of a protein nature, by injecting a purified hormone, x, from one species, say a pig, into a convenient laboratory animal such as a rabbit. The hormone from the pig will act as an antigen in the rabbit provided that it is sufficiently different in chemical structure from the rabbit's own hormone. Thus hormones such as insulin, growth hormone and gonadotrophins show species differences in chemical structure which permit them to act as antigens. Other smaller hormone molecules such as adrenaline, noradrenaline, oestradiol do not show such chemical variation between species and do not act as antigens. Provided that the hormone is antigenic, one can later separate from the plasma of the rabbit an antibody which the rabbit has manufactured against the pig antigen. This antibody we can refer to as anti-pig x antibody. If the antigen and antibody are mixed together they unite because of the affinity of the antibody molecule for those of the antigen x.

Antibody + Antigen = antigen–antibody complex
(anti-pig$_x$)   (pig$_x$)

It is possible to attach a fluorescent compound F to the extracted anti-pig$_x$ antibody, making a fluorescent antibody (F anti-pig$_x$). If this fluorescent antibody is applied to a section of pig tissue which contains the antigen x (in this case a hormone), then one can observe the distribution of the hormone by noting the distribution of the fluorescence. The fluorescent antibody combines with the hormonal antigen and is fixed as the fluorescent antigen–antibody complex.

$$F \text{ anti-pig}_x + \text{pig}_x = \text{fluorescent antigen–antibody complex}$$

Since the antigen–antibody union is highly specific, then fluorescence appears only in those parts of the tissue which contain the hormone. Using this technique it has, for example, been possible to localize the production of the hormone insulin to particular cells called β cells in the islets of Langerhans of the pancreas, since only these cells show fluorescence when a fluorescent antibody to insulin is applied to sections of pancreatic tissue.

Hormones can be converted into fluorescent compounds by methods other than the fluorescent antibody technique. It is now possible to localize the hormones adrenaline and noradrenaline in tissues by exposing freeze-dried tissue to formaldehyde. Subsequent examination of the tissue sections with the fluorescence microscope shows the position of the hormones which now exhibit a bright green fluorescence.

Autoradiography

This technique of 'self photography' has been used not only to localize the distribution of hormones in the body but also to obtain information on the sites of synthesis and the dynamics of secretion. The technique involves administration to an animal of a hormone containing one or more radioactive atoms in its molecular structure.

A variety of hormones such as thyroxine, testosterone, cortisone and oestradiol have been synthesized so as to contain radioactive atoms such as I^{131}, P^{32}, S^{35}, C^{14}, H^3. Alternatively, one can administer radioactive materials which the animals' own metabolic processes can turn into radioactive hormones. After an appropriate interval of time, tissue samples are removed from the animal and thin sections are prepared. These sections are placed in contact with a photographic emulsion which is sensitive to the radiations emitted by the radioactive hormone. The section is left in contact with the film until the emitted radiation has had time to affect the film. This may take days or weeks, depending on the kind and amount of radiation being emitted. When the film is developed, darkened areas represent the areas which have been affected by the radiation coming from the hormones. The section of tissue is still in position on the film and by examining the two together under the microscope and focusing up and down it is possible to see the part of the section which corresponds with the darkened areas of the film. The accuracy of the location of hormones, using the autoradiographic technique, is limited by various factors. For example, there is a limit set to the degree of resolution by the size of the grains of the emulsion on the film. Also the radiations are emitted in all directions from the tissue section, not just downwards at right-angles to the film, and this results in the areas of blackening on the film being larger than the corresponding sources of radiation on the section. Great improvements in film emulsions now allow a high degree of resolution, up to 0.1 μ in the case of hormones labelled with tritium (H^3). This allows the localization of radioactive sources within the cell, in such places as the chromosomes or the nucleolus. The best resolution is obtained with tritiated molecules on account of the weak energy of the β particles which they emit. Another advantage of using tritium is that its half-life period is 12 years and this means that there is a negligible amount of decay during the period of the experiment.

A good example of the use of the autoradiographic technique is seen in the investigation of the secretory activity of the hypothalamic region of the brain. As early as 1928, classical staining techniques showed the presence of secretory granules in neurones of certain regions of the hypothalamus, called the supra-optic and paraventricular nuclei. The axons of the neurones lying in these regions pass down to the posterior lobe of the pituitary gland. Secretory droplets were demonstrated in these axons and they were seen to accumulate in the posterior pituitary. From this histological evidence the view developed that neurosecretory material passes down the axons of neurones situated in the hypothalamus to the pituitary. The nerve terminals in the posterior pituitary are thought to liberate the polypeptide hormones, oxytocin and vasopressin. These ideas were confirmed and the dynamics of the process were made clear by the use of autoradiographic techniques using S^{35}-labelled amino acids. The hormones, oxytocin and vasopressin are polypeptides containing the amino-acids cysteine (p. 6). Radioactive sulphur-containing cysteine was injected into the cerebral ventricles and the animals were subsequently killed at various time intervals. Sections of the brain and pituitary were prepared and autoradiographed. Examination of the autoradiographs showed an early appearance of radioactivity in the hypothalamus (fig. 5), followed later by a gradual accumulation of radioactivity in the posterior lobe of the pituitary. The delay in the appearance of radioactivity in the posterior lobe is an indication of the time needed for the endocrine secretion of the neurones in the hypothalamus to migrate down the axons to the pituitary gland.

Using autoradiographic techniques one can follow the uptake of radioactive hormones by various organs, after injection of the hormones into the body. When tritiated oestrogens are injected into rats, some organs, e.g. anterior pituitary, uterus, vagina and certain 'sexual areas' of the brain take up the hormones selectively and bind them for much longer periods of time than do other non-target organs for oestrogen. In addition to tissue localization of hormones, one can in

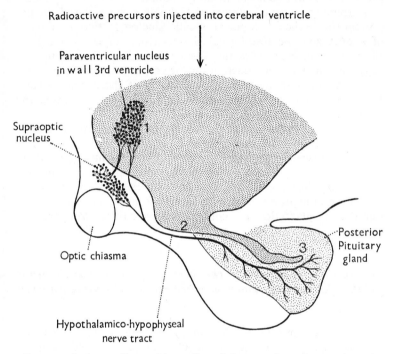

Radioactive precursors injected into cerebral ventricle

Paraventricular nucleus in wall 3rd ventricle

Supraoptic nucleus

Optic chiasma

Hypothalamico-hypophyseal nerve tract

Posterior Pituitary gland

FIGURE 5. Autoradiographic studies of the secretion of oxytocin and vasopressin. Vertical section of hypothalamus and pituitary gland showing the nerve tract linking the pre-optic and paraventricular hypothalamic nuclei and the posterior pituitary gland and the order in which radioactivity appears after the injection of radioactive precursors of oxytocin and vasopressin into the cerebral ventricle.

some cases locate cells containing enzymes which are concerned solely with the metabolism of the hormone. We can then infer that the hormone is also likely to occur there. A good example of the use of this technique is seen in investigations of the distribution of the hormone acetylcholine. Cholinergic nerve fibres, i.e. those releasing the hormone acetylcholine, can be detected by locating the enzyme cholinesterase, which is concerned with the degradation of acetylcholine.

Tissues sections are incubated with acetylthiocholine which serves as an artificial substrate for the enzyme cholinesterase. When the enzyme attacks this substrate, black sulphide compounds are formed and these show up as black areas on the section corresponding to the position in which the enzyme cholinesterase occurs. These same areas probably contain the hormone acetylcholine.

Morphological and histological techniques have played an important rôle in the development of endocrinology. Often the absence of a duct and the histological characteristics of the tissue have suggested an endocrine function for the tissue. Thus Prenant in 1898 suggested an endocrine function for the corpus luteum of the ovary, basing his opinion solely on the histology of the corpus luteum. Fraenkel (1903) and others, later confirmed this view using the experimental approach.

The morphological approach alone is, of course, limited and can indeed suggest investigations which may turn out to be fruitless. For many years several structures, including the thymus and spleen, which have the appearance of endocrine glands, have been investigated as possible sources of hormones without showing them to be so. Further, the presence of a duct leading from a gland is no indication that the gland does not contain endocrine tissue, e.g. pancreas, testis.

The experimental approach.—The effect of ablation of endocrine tissue and the effects of transplantation, organ extracts or purified hormones

Early work

The first experimental studies in endocrinology were made by Addison, Bertholt and Brown-Séquard in the mid-19th century. Addison (1849) described clearly a condition in which a variety of signs and symptoms were associated with destruction of the adrenal glands by disease. Brown-Séquard (1856) made pioneer experiments on the effects of removal of the adrenal glands from animals and confirmed that these glands were indeed essential for life. In 1849 Bertholt showed that transplantation of the testis into a previously castrated cockerel prevented the retrogressive changes in sexual character and behaviour which normally follow castration. He rightly concluded that the effects of the transplanted testis might be due to some substance secreted into the blood. Although these two important aspects of the experimental approach in endocrinology, viz. (i) the effects of removal of an endocrine organ, and (ii) the restorative effects of transplantation, were established early in the history of physiology, it was not until the early 20th century that work progressed rapidly, when the techniques of isolation and purification of the active principle of an endocrine gland became perfected. For example, in 1899 Oliver and Schafer studied the effects of extracts of the adrenal medulla on the blood pressure of an experimental animal. They found that intravenous injection of medullary extracts produced marked rises in blood pressure. These investigations excited great interest, which soon resulted in the isolation, identification and synthesis of the active principle, adrenaline.

Transplantation experiments

Transplantation of endocrine organs is a venerable technique in endocrine research, dating as it does from Berthold's work in 1849. If the symptoms which appear on removal of a gland are nullified by the subsequent transplantation of a normal gland into the animal, then this leads one to attribute an endocrine function to the gland. The transplanted gland can exert effects on the host by discharging chemical activators into the blood stream.

TRANSPLANTATION AND IMMUNITY
Most recognized endocrine glands can be successfully transplanted. The success of this technique is at odds with present-day views on immune responses to transplanted tissues. Organs and tissues transplanted from one individual to another are often totally or partially rejected by the host animal unless there is a very good measure of genetic similarity between donor and host, or unless the tolerance of the host is increased by damaging the immunological defence mechanisms of the host by drugs or X-ray treatment. The endocrinologist is particularly fortunate in that transplanted endocrine tissues do not elicit a strong immunological response from the host. However, transplanted endocrine tissues never seem to have been widely used in the management of human diseases caused by malfunction of endocrine glands.

Experimentally, perhaps the most useful type of transplantation is that in which the organ is transplanted from its normal position to another site in the same body, since this avoids the immunological problems which would be involved if the transplanted organ were placed in a different body. In any transplantation experiment the only connection between the transplant and the rest of the body is via the blood vessels which grow into the transplant from the surrounding tissues.

Cross-circulation experiments

If endocrine glands exert their effects by hormones liberated into the circulating blood, then it should be possible to show effects in a second animal, which is connected to the first animal only by its blood stream.

LOEWI'S CLASSICAL EXPERIMENT

Loewi's brilliant research in 1924, which provided the first proof that nerve impulses produce their effects on tissues by the release of a specific chemical substance, give us the archetype of the cross-circulation experiment. Loewi worked with a frog heart perfused with a physiological saline solution. When he electrically stimulated the vagus nerve to the heart it produced a marked slowing of the rate of beating. When the fluid perfusing this heart was allowed to come into contact with a second heart, this also showed a slower rate of beating. He postulated that a chemical substance which he called 'vagustoff' was liberated when the vagus nerve was stimulated (see fig. 6).

AN ENQUIRY INTO ALDOSTERONE USING CROSS-CIRCULATION TECHNIQUES

Cross-circulation experiments of various types have been successfully used in a variety of investigations which have demonstrated the existence of endocrine mechanisms. It is known that during adaptation to lack of sodium in the diet or to excessive loss of sodium from the body, e.g. in sweating, the adrenal cortex produces increasing quantities of the hormone aldosterone. Aldosterone acts on the distal convoluted tubule of the nephron in the kidney where it stimulates the absorption of increased quantities of sodium from the glomerular filtrate, thus helping to maintain the sodium balance of the body. It was not known whether the adrenal cortex could itself 'sense' the changing sodium concentration of the blood. To facilitate the experimental study of this problem, the adrenal glands of sheep were transplanted into a skin pouch of the neck, anastomosing the adrenal vein with the jugular vein and the adrenal artery with the carotoid artery. In such an accessible site it was possible, without the complications of anaesthesia and surgery, to infuse solutions of known sodium content into the adrenal artery and at the same time collect blood samples from the adrenal vein for analysis of its aldosterone content. Using this technique it was found that a fall in the sodium content of the blood in the adrenal artery did not lead to an increased output of aldosterone, thus indicating that the adrenal cortex is not sensitive to changes in the sodium concentration of the blood.

In other experiments blood was cross-circulated from sheep which had been adrenalectomized and subjected to sodium depletion to normal sheep. The recipient sheep were caused to secrete increasing quantities of aldosterone. This experiment suggests that during the sodium depletion of the donor sheep a hormone appeared in the

blood, which on cross circulation stimulated the adrenal cortex of the recipient sheep to secrete aldosterone.

The next step in the investigation was to remove various tissues of the body in an attempt to determine the source of this hormone. If removal of the pituitary gland, for example, interfered with the response of the animal to sodium deprivation, then this might suggest that the pituitary gland was capable of detecting a deficiency of sodium in the blood. It might also suggest that the pituitary was capable of producing a hormone, activating the adrenal cortex to secrete aldosterone. Removal of various tissues, e.g. pituitary gland, the central nervous system anterior to the mid-collicular region and the pineal gland, have all failed to disturb the mechanism of adaptation to sodium deprivation. We now know the source of one hormone which appears in the blood and activates the adrenal cortex when the amount of sodium in the blood falls (see p. 86).

CROSS CIRCULATION TO AN ISOLATED ORGAN

A modification of the cross-circulation technique was used by Copp (1962) in the investigation of the rôle of the parathyroid gland in the regulation of the concentration of the calcium ions in the blood. The classical parathyroid hormone, parathormone, had been known since 1925 when Collip showed that the injection of an extract of parathyroid tissue would cause a rise in the calcium level of the blood. This it does by mobilizing the calcium stores of bone. By this means the calcium content of the blood can be maintained at normal levels, even when the diet is temporarily deficient in calcium ions. Various pieces of evidence had suggested that in addition to parathormone, the parathyroid gland produced another factor which could reduce the levels of calcium in the blood. The interplay of the two hormones, one controlling the rise in concentration of calcium (parathormone) and the other controlling a fall in calcium concentration (a hypocalcaemic factor), would readily explain the exact control of the amount of calcium in the plasma (see p. 68). This is difficult to explain if only one hormone is involved. Copp was able to demonstrate the presence of a hypocalcaemic factor in the parathyroids of a dog in the following way. Parathyroid glands were isolated from the dog's body and placed in a warm moist chamber. Dog's blood was perfused nto the gland via the artery and the venous blood leaving the gland was allowed to drip into the jugular vein of another dog. Now the level of calcium in the blood perfusing the isolated parathyroid gland was raised 2 mg % by the addition of calcium chloride to the perfusing blood. When this was done there was a rapid fall in the calcium level of the plasma of the dog receiving blood from the perfused isolated glands. These results were taken to indicate the presence of a rapidly acting hypocalcaemic factor which is produced by the para-

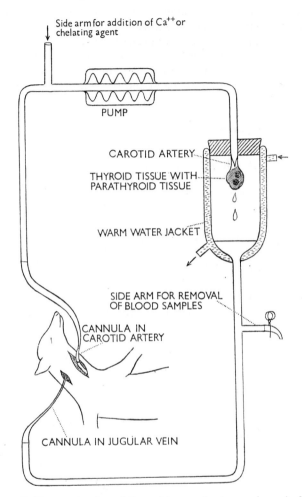

FIGURE 7. Demonstration of the existence of a hypocalcaemic factor in perfusates of isolated parathyroid and thyroid tissue (after Copp, 1961).

thyroid gland and controls the fall of calcium in the blood (fig. 7). Modern ideas on this topic are discussed more fully on p. 68.

PARABIOSIS

Another type of cross-circulation experiment is one in which a permanent union is established between two animals. This is known as parabiosis. For this type of work one must choose young animals of a closely inbred strain, to avoid immunological problems arising between the two individuals. The two animals are joined surgically over a large area of the lateral surface of the body. Tissue union of the two is established during healing and a cross circulation develops. Parabiotic union has proved a useful method in investigating the relationships between the pituitary gland and the gonads (p. 178).

Further investigations

Using the variety of techniques which have already been described it is possible to establish the presence of endocrine mechanisms in particular processes. The establishment of the presence of such a mechanism now poses new questions. What is the chemical nature of the hormone(s) involved? What is the chemical structure of the hormone? What is its molecular structure and can it be purified in amounts sufficiently large to enable an experimental programme to be undertaken? Can the hormone be synthesized? In what cellular and subcellular structures is the hormone produced and stored? How is the production and release of the hormone controlled? What is the significance of the hormone in the normal physiology of the animal? And finally, how in fact does the hormone produce its effects on its target cells?

These are enormous questions which can only be answered by the co-operation of the physiologist, the chemist, physicist, biochemist, immunologist and the geneticist.

Separation and isolation of hormones

The problems that arise in the initial extraction of the active hormone from the secretory tissue may be enormous. The quantities of hormone stored may be small and extraction therefore involves the use of large quantities of raw materials. Literally tons of raw material may be necessary to obtain a few milligrammes of material for the analysts to work upon. The first step in extracting the hormones from the tissue extract is the separation of the active material from the very large number of other biologically active substances, such as histamine, nucleic acids, acetylcholine, 5-hydroxytryptamine, adrenaline, noradrenaline, etc. The use of appropriate solvents for extraction may eliminate many of the unwanted materials. The extract of tissue may then be exposed to a variety of processes in an attempt to separate the various constituents, which can then be tested for biological activity. The kind of processes involved in this stage of separation are chromatography, electrophoresis, ultracentrifugation and counter-current distribution.

Determination of the chemical nature of the hormone

Having obtained a relatively pure sample of active material, biophysical and biochemical methods can now be employed to determine the molecular size, shape and constitution. Some of the techniques here depend on the different diffusion and sedimentation rates of the molecules or upon their absorption of light and X-rays. Substances with chemical bonds such as $C-H$, $C = O$, $C-O$, etc., show particular absorption regions in the infra-red zone of the spectrum. Series of atoms with alternating double and single bonds such as $-C = C-C = C-$ absorb at wavelengths of light which increase in proportion to the length of the series, the shorter series absorbing in the ultra-violet region. Biologically important molecules such as nucleic acids and proteins may contain short series of alternating bonds in parts of their molecules. The absorption of infra-red and ultra-violet light by solutions of such molecules is in proportion to the concentration of the molecules. Thus an instrument such as a colorimeter, or spectrophotometer, which measures light absorption can be used to determine the concentration of the materials present. These instruments can also be used to identify the material present if pure reference compounds are available for comparison. These techniques have not directly given us any information about the size and shape of the molecule. The absorption of X-rays can be used for this. When X-rays impinge upon a substance, most of the rays pass through the matter without deflection if the matter is not too thick. Some of the rays are, however, scattered, and the particular pattern of scattering depends upon the pattern of the atoms which make up the molecules of the substance. In some cases it is possible to compute the structure of the scattering material from the pattern of scattering of the X-rays. This technique was first used by Bernal over 30 years ago. He took crystals of oestrone (a female sex hormone) and examined the diffraction patterns which were obtained when X-rays were passed through the crystals. The results were recorded on film. This information gave him some idea of the size of the sub-units in the crystal, and he formed the impression that the molecules of oestrone were long and flat.

Storage of hormones in tissues

The quantities of hormones which are stored in tissue vary greatly from one endocrine organ to another. The human adrenal cortex contains very small quantities of aldosterone, but the human pancreas contains 200 international units, which is more than a lethal dose for a man if it were injected into him all at once as a free hormone. Obviously some hormones are therefore stored in an inactive or

B

inaccessible form in the secreting cell. The stores are often in the form of distinct granules which can be observed by the light or electron microscope. Hormones which have been demonstrated in the granular fraction of tissue homogenates include, acetylcholine, 5-hydroxy-tryptamine, histamine, peptide hormones of the hypothalamus and posterior pituitary, noradrenaline, adrenaline and insulin.

Synthesis of hormones

The synthesis of a hormone may be a very important step in the history of the study of a particular hormone. Not only does it provide quantities of the hormone available for the treatment of human deficiency diseases but also it facilitates the investigation of the physiological and pharmacological effects of the hormone. When a new hormone becomes available in a chemically pure form in reasonable quantities, research on this hormone accelerates. The hormone is injected into animals of many species, both vertebrate and invertebrate, into the blood stream, into the cerebral ventricles and is applied to the skin. It may be added to tissues which are maintained alive outside the body, or to a suspension of cells or subcellular particles, such as mitochondria, nuclei or ribosomes. The hormone may be added to enzyme systems. This type of work provides the information we need to answer, if only partially, the question 'how do hormones act?'.

The hormones which have been fully synthesized chemically include the polypeptide hormones oxytocin and vasopressin, gastrin, the adrenal steroids cortisol and cortisone, adrenaline and noradrenaline and various sex hormones. The complex nature of many hormones, in particular the proteins, have so far proved too difficult for synthesis and therefore supplies of hormones such as pituitary gonadotrophins and growth hormone, and of pancreatic insulin, have still to be obtained from biological sources. The amino-acid composition and sequence of these hormones varies from species to species and it is important to remember this when one uses a protein hormone in one species which is obtained from a different species. Supplies of follicle stimulating hormone (FSH) for the treatment of some cases of infertility in man have to be laboriously obtained by extraction from human pituitary glands removed after death.

3

The Mode of Action of Hormones

Ever since the 19th century endocrinologists have been concerned with the rôle of endocrine mechanisms in the production of a highly integrated organism adapted to its surrounding environment. More recently, with the advent of cellular and molecular biology, increasing attention has been given to the problem of how individual hormones exert their effects in the organism.

The ultimate interaction of a hormone with its target cell is at the molecular level. We shall be considering examples of such interactions at various levels in the cell, including the cell membrane, enzyme systems, intracellular organelles such as ribosomes and the genetic material in the cell nucleus.

The action of insulin on cell membranes

The nature, storage and release of insulin

Before considering the evidence for the idea that insulin produces some of its effects in some cells by means of an action on the cell membrane let us first look at some of the known facts about this hormone.

That the pancreas had functions other than digestive ones was discovered as early as 1899 by Mering and Minkowski, when they found that the removal of the pancreas from a dog resulted in various metabolic derangements which are also seen in the disease of diabetes mellitus in man. They found that tying off the pancreatic duct in the dog did not cause these disorders, thus indicating that the disorders did not arise from a lack of the digestive juices which normally flow down the duct. Later work showed that the development of diabetes in the dog after removal of the pancreas could be delayed by transplanting a part of the pancreas under the skin. This showed that the pancreas had an internal secretion as well as the external secretion. However, the many efforts which were made to extract the active principle from pancreatic tissue proved unsuccessful until Banting and Best in 1921 realized that the lack of biological activity shown by the pancreatic extracts was due to the presence of digestive enzymes in the tissue. They were able to overcome this problem by first ligating the pancreatic duct of dogs, which produces atrophy of the exocrine tissue producing the digestive enzymes. Some months later the dogs were killed and extracts were made of the shrunken pancreas. These extracts were found to have marked biological activity. The active principle of the extracts is a protein called insulin.

Insulin was the first protein to have its structure elucidated. Sanger showed that the molecule of insulin is composed of two polypeptide chains, an A chain containing twenty-one amino acids and a B chain containing thirty amino acids. The two chains are linked together by two sulphydryl groups belonging to cysteine residues in the chains. The complete molecule of insulin does not seem to be essential for biological activity and many different peptides with insulin-like activity have been derived from insulin. Biological activity is, however, lost if the disulphide bridges between the two chains are broken.

The hormone is synthesized and stored in certain cells of the islets of Langerhans in the pancreas (plate 1). These cells are called beta cells and the hormone can be localized here using the fluorescent antibody technique (see p. 14). Within the cell insulin appears to be located inside granules. The most potent stimulus to elicit the secretion of insulin by the beta cells is a rise in the concentration of glucose in the blood. This effect of glucose has been shown in the whole animal, in the isolated perfused pancreas, and in pieces of pancreas suspended in physiological saline solution. After release from the beta cell the hormone enters the blood stream, where it combines with several of the plasma proteins. In this form the hormone may be protected from a variety of insulin-destroying mechanisms which exist in the body, and in this form it is thus capable of acting as a circulating store. To reach the site of action, insulin has to leave the blood, presumably to become bound to the cells on which it acts. This idea of the binding of insulin by cells arose from observations made on isolated tissues. It was found that the action of insulin persisted in tissues which have been briefly exposed to the hormone and then washed free of the hormone. It has been suggested that insulin binding by tissues involves an interchange reaction between the thiol groups on the cell surface and a disulphide bond in insulin. This idea is supported by the fact that the application of poisons for the thiol groups (N-ethyl-maleimide) to tissues suppresses the action of insulin on these tissues. Further, such substances as thioglycolate and cycteine which can disrupt disulphide links can cause the release of insulin from tissues which have previously bound the hormone.

The actions of insulin

Administration of insulin to intact animals or to isolated tissues produces a variety of effects. In muscle tissue there is an increased rate of uptake of glucose by the cells and glycogen is formed. In adipose tissue there is an increased rate of formation of fatty acids and triglycerides, which is due to an increased rate of uptake of glucose by the fat cells. In a variety of tissues there is also an increase in the rate of synthesis of protein. This last effect is not due to an increased rate of uptake of glucose by the cells, because it still occurs in isolated tissues when glucose is absent from the physiological saline.

Because of the increased rate of uptake of glucose by the various tissues a prominent effect of insulin, when it is injected into the whole animal, is a fall in the concentration of glucose in the blood. Unlike some other tissues of the body the central nervous system is dependent almost exclusively on glucose as a source of energy so that disturbances of function of the nervous system may appear if the glucose concentration in the blood falls to a critical level. These disturbances may appear in the form of convulsions and eventually coma.

Mode of action of insulin

According to our present state of understanding these various effects of insulin cannot be explained by a single mechanism or site of action.

The action of insulin on protein metabolism is regarded as an effect of the hormone at the level of the gene. This view is based on the evidence that insulin produces marked changes in the amount of RNA in the cell, an effect which can be blocked by the agent actinomycin (p. 42). The effect of insulin on the uptake of glucose by muscle and fat cells, for reasons which will be described below, is regarded as being the result of an activation of the cell membrane transport system for glucose, an effect which can be seen at low temperatures and one which is not blocked by actinomycin.

Evidence for the site of action of insulin on carbohydrate and fat metabolism

THE UPTAKE AND METABOLISM OF GLUCOSE BY CELLS
Before looking at the evidence for the site of action of insulin let us

first consider some of the factors which are involved in the uptake and metabolism of glucose by cells.

The passage of glucose from the circulating blood into the interior of cells involves movement of the molecules of glucose across two

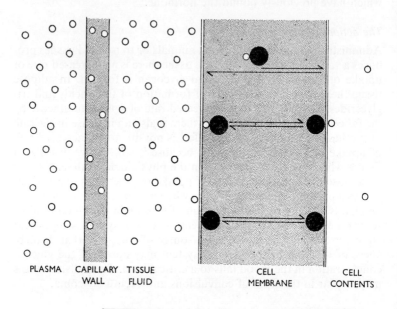

PLASMA   CAPILLARY   TISSUE   CELL   CELL
             WALL   FLUID   MEMBRANE   CONTENTS

GLUCOSE CONCENTRATION GRADIENT

FIGURE 8. Diagram showing the membranes separating glucose in plasma from the cell contents. The capillary wall presents no barrier to exchange of glucose between plasma and tissue fluid and the concentration of glucose ○ is the same in the two spaces. The cell membrane is a barrier to the free movement of glucose from tissue fluid into the cells and glucose molecules can only gain access by combination with a carrier molecule ● which oscillates across the cell membrane. This scheme particularly applies to muscle and fat cells. Some cells, e.g. liver and brain, do not show such a membrane barrier to the movement of glucose.

membranes, first the capillary wall which gives access to the interstitial fluid of tissues, and second, the cell membrane which give access to the interior of the cell. The transfer of glucose from blood to the interstitial fluid is regarded as taking place under the influence of simple physical forces and the capillary wall appears to present no barrier to the diffusion of glucose molecules along a concentration gradient. The concentration of glucose in the interstitial fluid of tissues rapidly comes into equilibrium with the glucose of blood when this is artificially altered. However, the transfer of glucose from the interstitial fluid into cells appears to depend on a specific mechanism located in the cell membrane, a system of active transport. It is suggested that the membrane contains a substance called a carrier which is capable of entering into a reversible combination with glucose. It is supposed that the carrier is capable of oscillating or diffusing across the cell membrane and that association with the carrier at the outer surface of the cell membrane and dissociation from it at the internal surface of the membrane is the only way in which glucose can cross the cell membrane (fig. 8). This transport mechanism shows Michaelis Menton kinetics (i.e. its behaviour is predictable from the law of mass action), stereospecificity (i.e. only sugars with particular steric configurations are transported), and competition between different sugars for the transport mechanism.

Once inside the cell the glucose is phosphorylated by reaction with ATP under the influence of the enzyme hexokinase to form glucose-6-phosphate. Glucose-6-phosphate is in equilibrium with other hexose monophosphates in the cell, glucose-1-phosphate and fructose-6-phosphate. The hexose monophosphates may be utilized by way of the glycolytic pathway, pentose phosphate pathway, uronic acid pathway or by way of the glycogen path.

There have been two main views on the way that insulin facilitates the uptake of glucose by some cells.

The action of insulin on the intracellular enzyme hexokinase

Because insulin influences several routes of glucose utilization by cells, e.g. the synthesis of fat and the synthesis of glycogen, it has been argued that the hormone must act on some initial stage of the utilization of glucose. This stage is the phosphorylation of glucose by ATP in the presence of the enzyme hexokinase. In 1947 Cori provided evidence for this viewpoint. Using tissue extracts he found that the addition of insulin could abolish the delay in the activity of hexokinase. (Usually when these investigations are made on tissue extracts the hexokinase does not act immediately. The time taken before activity starts is called the lag or the delay.) It was thought that insulin produced this effect by removing some inhibiting influence from the system rather than by a direct influence of the hormone on the enzyme hexokinase. However, no other workers have yet been able to repeat Cori's observations, and indeed there is no other experimental evidence to support the view that insulin acts on the hexokinase reaction. If insulin acted on the hexokinase reaction, then the application of the hormone to tissues should result in the disppearance of free glucose from the interior of the cells. In fact the opposite occurs and one can detect the presence of increased amounts of free glucose inside cells after treatment with insulin.

Action of insulin on membrane transport mechanisms for glucose

In 1939 Lundsgaard obtained experimental evidence which led him to the conclusion that insulin influences the mechanism which transfers glucose from the extracellular fluid to the interior of cells. His experiments were performed on cats whose viscera, including the pancreas, had been removed. His experimental animals could not produce insulin, of course. He injected glucose into a vein and measured the glucose content of the muscle tissue at various concentrations of glucose in the blood. The amount of glucose in the muscle tissue did not exceed the amount of glucose which he calculated was present in the muscle tissue fluid. This amount can be calculated by knowing the volume of the interstitial fluid of tissue and the concentration of glucose in the plasma, since the glucose concentration of interstitial fluid is identical with that of plasma. He came to the conclusion that the concentration of glucose in the muscle cells must be very low and that the transfer of glucose from the extracellular fluid into the interior of the cell is a very important limiting factor in the utilization of glucose by cells. On injecting insulin into the animal he found an increased concentration of glucose in muscle tissue, which he attributed to an uptake of glucose by cells.

Lundsgaard's view of the action of insulin was subsequently supported by the experimental evidence obtained by Levine (1949). Levine worked with dogs from which the abdominal viscera and kidneys had been removed. In these animals he studied the effect of insulin on the distribution of galactose, a sugar which is not metabolized by the body. This experimental situation avoided some of the uncertainties about the fate of a sugar injected into the body since the sugar could be neither metabolized nor excreted by the animals. Levine studied the volume of distribution of the sugar following its injection into the blood stream. Using the relationship between the amount of a substance known to be added to an unknown volume of fluid and the final concentration of the substance in the fluid one can calculate the volume of the fluid.

$$V = \frac{A}{C}$$

where V is the volume, A the amount of substance added, and $C =$ final concentration of substance.

After the injection of a known amount of galactose into the circulation there is a fall in the concentration of galactose in blood which is due to the diffusion of galactose into other fluid compartments of the body such as interstitial fluid and intracellular fluid. Levine found that after injection of galactose into the blood the final volume of distribution of galactose, calculated from the known amount injected and the blood concentration after sufficient time had elapsed for

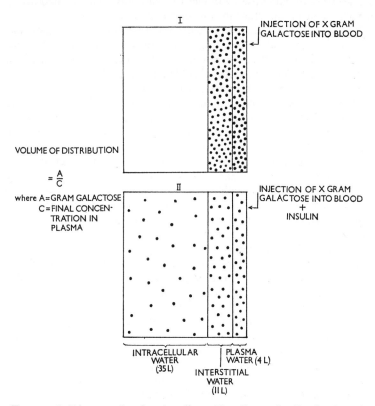

FIGURE 9. Diagram showing the effect of insulin on the distribution of a sugar (galactose) in the various fluid compartments of the body. Note the different sizes of the fluid compartments. The number of litres for each compartment refers to average figures for man; in I galactose is injected into a vein and becomes, in the absence of insulin, distributed predominantly in extracellular fluids (plasma and interstitial water). The volume of distribution is given by the formula; in II the same amount of galactose is injected but with insulin. The final concentration of galactose in plasma is less than in I, i.e. the volume of distribution is greater. This is because insulin promotes the transport of galactose into the cells. Note that the concentration of galactose in cell water does not equal that in extracellular fluids.

galactose in blood to come into equilibrium with galactose in other fluid compartments, was that of the extracellular fluid volume (i.e. plasma volume + interstitial fluid volume). When insulin was injected the concentration of galactose in blood decreased and thus the calculated volume of distribution increased. In the absence of the metabolism of galactose by tissues this can only be accounted for by a movement of galactose into intracellular fluid (fig. 9).

This work had been confirmed by many workers using various isolated organs and tissues. It can be readily shown that after treatment with insulin there is a rise in the concentration of a variety of metabolizable and non-metabolizable sugars inside cells when these sugars are present in the perfusing or suspension medium.

It has been suggested that the insulin molecule may enter into a reversible combination with the membrane carrier molecule, so altering the properties of the carrier molecule to increase the efficiency of transport. One way in which this might occur is by a reduction of the affinity of the carrier molecule for glucose, i.e. a reduction in the tendency for the carrier and glucose to combine. Reduction of the affinity between glucose and the carrier molecule would permit a more ready release of glucose from the carrier at the internal surface of the cell membrane. It would, of course, also reduce the amount of glucose picked up at the outer surface of the cell membrane. It may seem paradoxical that a reduction in the affinity of the carrier for glucose could increase the rate of glucose transport. If a carrier had infinite affinity for glucose, then little or no glucose would be released into the cell, however low the concentration of glucose in the cell. Conversely, a carrier with a low affinity for glucose would also result in minimal glucose transport since little glucose would be accepted at the outer surface of the cell membrane. Obviously there is an optimal affinity of the carrier for the glucose molecule for efficient transport of glucose. Certainly little glucose is transported across the membrane of many cells in the absence of insulin, and the hormone may increase the transport of glucose by modifying the properties of the carrier molecule to bring its affinity for glucose into an optimal range.

The action of the hormone glucagon on the intracellular enzyme phosphorylase

The hormone glucagon

The injection of some preparations of insulin produces a temporary rise in the concentration of glucose in blood which precedes the typical fall in concentration of glucose. The initial effect is due to contamination of insulin with a second pancreatic hormone which has been isolated and crystallized. This hormone is glucagon, the molecule of which is a straight-chain polypeptide containing twenty-nine amino-acid residues.

There is considerable evidence which indicates that this hormone is produced by the alpha cells of the Islets of Langerhans. Like insulin the hormone can be extracted from the pancreas after the exocrine component of the gland has been destroyed following ligation of the pancreatic duct. Further, one can specifically destroy the beta cells of the Islets of Langerhans following duct ligation by the intravenous injection of a substance alloxan which is highly toxic to beta cells. In such animals without either exocrine tissue or beta cells in the pancreas one can still extract the hormone glucagon from the pancreas.

Glucagon produces a variety of effects when administered to animals. One marked effect, the mode of action of which will be discussed here, is a rise in the concentration of glucose in blood.

THE BREAKDOWN OF LIVER GLYCOGEN AND THE ACTION OF GLUCAGON

Glycogen is a polysaccharide in which glycosyl units are joined together by α–1,4 links to form straight chains. Straight chains are further joined to one another by α–1,6 links to produce a branching structure (fig. 10). Units of glucose can be added to or removed from the branching molecule which has then a very variable molecular weight.

The breakdown of the glycogen molecule in the liver, which results in the formation of free glucose, takes place in three stages. In the first stage glycogen reacts with inorganic phosphate in the presence of the enzyme phosphorylase. This reaction fissures the α–1,4 links between glycosyl units of the straight chains and liberates glucose-1-phosphate. This process proceeds along the straight chains of the molecule until an α–1,6 link of a branching point is reached. This link is broken under the influence of another enzyme, debranching enzyme, which then permits the enzyme phosphorylase to continue its action along the straight chain. The glucose-1-phosphate which is liberated by the action of phosphorylase can then be converted to glucose-6-phosphate by the enzyme phosphoglucomutase. Free glucose can be derived from glucose-6-phosphate by means of an irreversible reaction catalysed by the enzyme phosphatase. Although many other tissues, e.g. cardiac muscle, skeletal muscle and brain contain glycogen they do not contain this phosphatase. These tissues, then, cannot provide glucose for the general body economy and the glucose-6-phosphate derived from glycogen is utilized for the tissues' own metabolic needs.

The rate of glucose production from glycogen in the liver depends upon the amount of the enzyme phosphorylase in liver cells. The amount of phosphorylase varies from moment to moment and depends upon the rate of inactivation and reactivation of the enzyme. Phosphorylase is inactivated by a specific enzyme. Reactivation of phosphorylase depends upon rephosphorylation by ATP in the presence of an enzyme dephosphorylase kinase. This enzyme must be

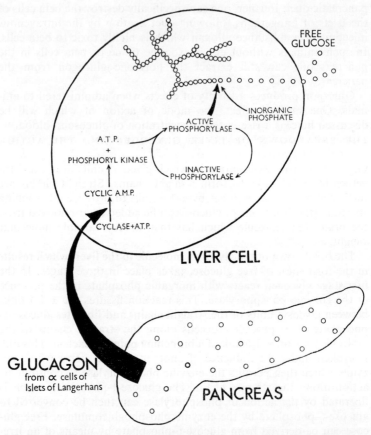

FIGURE 10. Scheme showing mode of action of the hormone glucagon.

activated by cyclic adenosine monophosphate, the amount of which depends upon the rates of production and destruction of the nucleotide. The nucleotide is produced in the presence of various factors, including ATP, the enzyme cyclase and magnesium ions. It is at this level that the hormone glucagon acts. Glucagon stimulates the rate of formation of cyclic AMP, which, by activating phosphorylase kinase causes a regeneration of the enzyme phosphorylase (fig. 10). Following the injection of glucagon into the whole animal there is an increased amount of active phosphorylase in liver cells and a decrease in glycogen content due to the breakdown of glycogen and the liberation of free glucose into the circulating blood.

Various other hormones, including adrenaline and ACTH, have been shown to exert their effects on the concentration of glucose in blood in a similar way.

The significance of these effects of insulin and glucagon is discussed in chapter 6.

The action of hormones at the level of the gene.—I. Cortisone

INTRODUCTION
The importance of the adrenal glands in the life of mammals has been appreciated since the mid-19th century from the observations of Addison and Brown-Séquard. However, it has taken over a hundred years of scientific investigation to provide partial explanations of how the hormones of the adrenal cortex produce their effects. Physiologists and biochemists for many years have seemed to be obsessed with the idea that the regulation of cellular metabolism by hormones must be due to an effect of the hormone either on the cell membrane or on intracellular enzymes. We have already discussed one effect of insulin at the level of the cell membrane, and one could list several hormones which do appear to act at this site in the cell. However, studies of the effects of hormones on enzyme systems, with the exception of a few clear examples, have proved to be rather unrewarding in spite of extensive research in this field. As late as 1962, Professor Bush in a review article on the action of adrenal steroids at the cellular level had to write, 'Despite much intensive work on the problem, the actions of all classes of steroid hormones at the molecular level remain unexplained.' He came to the conclusion that this problem remains as one of the major challenges of physiology and biochemistry.

Advance has been made so rapidly that in the same year that Bush had to come to this conclusion, Kenney was able to demonstrate that cortisone caused the *de novo* synthesis of a particular enzyme in liver cells. And in the same year it was found that actinomycin D, a substance which can specifically block DNA-dependent RNA synthesis could block this action of cortisone on liver cells. These results pointed to a site of action of cortisone at the level of the nuclear chromatin, on the gene. Except for the ultimate question of how cortisone influences the gene we now have a fairly complete picture of how this hormone produces some of its physiological effects. But before considering further the mode of action of cortisone we must first examine the nature of the hormone and its physiological rôle in the organism.

Hormones of the adrenal cortex

The adrenal cortex (plate 2) shows three distinct zones when suitably prepared sections are examined under the microscope; an outer zona glomerulosa, a middle zona fasciculata and an inner zona reticularis. Functionally, however, there are two zones which differ in the hormones they produce, in their significance to the organism and in their mode of regulation. The outer glomerulosa, with its densely staining lipid-poor cells is concerned with the elaboration and release of the hormone aldosterone (p. 80). This hormone forms part of the mechanisms of control of body fluid and electrolytes. Unlike the zona fascicularis and reticularis the zona glomerulosa remains intact and functional after removal of the anterior pituitary gland from an animal. The mechanisms which regulate the activity of the glomerulosa are still the subject of research and debate.

The zona fasciculata and reticularis form a functional unit whose main function is the production of the steroid hormones cortisol (hydrocortisone) and cortiscosterone (cortisone) together with some sex steroids, e.g. weak androgens and possibly some oestrogens. The zona fasciculata with its lipid-rich cytoplasm stores the raw materials for the synthesis of these hormones. The denser staining zona reticularis is the actual biosynthetic zone, although under a strong stimulus, such as an injection of pituitary adreno-corticotrophic hormone, the two areas become unified into a single biosynthetic zone.

The hormones produced by the adrenal cortex are collectively referred to as corticoids. A corticoid is a substance produced by the adrenal cortex and possessing twenty-one carbon atoms and three more oxygen atoms (p. 12). The composition of the mixture of C21 steroids produced by the cortex differs from one species to another and even in a particular species the composition of the mixture changes with the physiological state of the animal. Cortisone and cortisol have very similar physiological effects and are included in the general term glucocorticoid, which emphasizes their rôle in the metabolism of carbohydrate. And it is the mode of action of this class of cortical hormones which we shall be discussing here.

Glucocorticoids have a short life after they are released from the adrenal cortex into the general circulation. The half life of these compounds is only four hours, thus ensuring the rapid elimination of potent hormones which are being continuously secreted by the adrenal cortex. They are inactivated in various ways. In the liver they are conjugated with glucuronic acid and reduced to their tetrahydroderivatives which are biologically inactive. Both free and conjugated steroids are rapidly eliminated from the blood in the kidney by glomerular filtration and they are only partially reabsorbed by the renal tubules.

PHYSIOLOGICAL ACTIONS OF GLUCOCORTICOID HORMONES
Glucocorticoids influence the following:

1. Carbohydrate metabolism.
2. Protein metabolism.
3. Lipid metabolism.
4. Responses to infection, injury, etc.
5. Mechanisms maintaining blood pressure.
6. Metabolism of fluid and electrolytes.

Following the removal of the adrenal glands from an animal the first abnormality to appear is a disturbance of fluid and electrolyte balance. There is an increased loss of sodium and water in the urine, such that the loss exceeds the dietary intake. This is due to a lack of the hormone aldosterone. By contrast physiological doses of glucocorticoid hormones have little effect on salt and water metabolism.

If the adrenalectomized animal is kept alive by allowing it to drink saline solution or if aldosterone is administered, then a second defect in metabolism appears which reflects the lack of glucocorticoid hormones.

GLUCOCORTICOIDS AND THE METABOLISM OF CARBO-
          HYDRATE, PROTEIN AND FAT
One clear defect of the adrenalectomized animal is an inability to maintain normal amounts of glucose in the circulating blood during periods of fasting. The concentration of glucose in blood may fall to such a level that disturbances of function of the central nervous system appear (p. 72) and hypoglycaemic convulsions may develop. In association with this defect there are poor stores of glycogen in the liver. The normal animal maintains a normal content of blood during fasting by utilizing body protein as a source of carbohydrate—a process which occurs in the liver (p. 73).

Contrasting with the effects of adrenalectomy are the results of administering large doses of glucocorticoid hormones to normal animals. This treatment causes an increase in the glycogen content of the liver and a rise in the concentration of glucose in the blood. There may be so much glucose in the glomerular filtrate in the kidneys that the amount exceeds the ability of the nephrons to reabsorb glucose, so that glucose appears in the urine. Glucocorticoid hormones can produce these effects even in the fasting animal, which indicates that the glucose must be coming from some endogenous source. In fact, in parallel with the increased rate of formation of glucose there is an increase in the urinary excretion of nitrogen. It is clear that the new carbohydrate appearing after administration of glucocorticoids is arising from the breakdown of protein, a process called gluconeogenesis. The liver is the target organ for this effect of corticoids. Because of this effect on protein metabolism corticoids are powerful inhibitors of growth.

The rise in the concentration of glucose in blood is not entirely explained by an increased rate of production of glucose by the liver. There is also evidence that there is a reduction of the rate at which

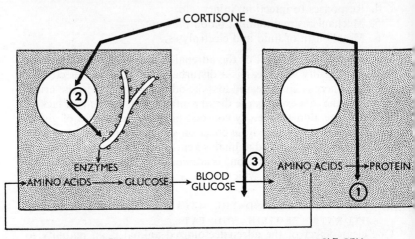

FIGURE 11. Scheme showing three actions of cortisone which account for the rise in concentration of glucose in blood. 1. In many body cells cortisone reduces the rate of protein synthesis. Amino acids are now available for gluconeogenesis in the liver. 2. In liver cells cortisone produces an increase in the amount of enzymes involved in gluconeogenesis. 3. Cortisone reduces the rate of uptake of glucose by many body cells. Glucose is thus conserved for the requirements of the C.N.S.

many tissues utilize glucose, and in particular a reduction in the synthesis of fat. The physiological significance of this dual effect of cortisone on carbohydrate metabolism, i.e. an increased rate of formation of glucose and a reduced rate of glucose utilization will be discussed further in chapter 6. Here we can say that this dual effect ensures that the central nervous system is guaranteed a continual supply of glucose even when there are inadequate amounts of carbohydrate in the diet or during periods of complete fasting. The actions of cortisone are summarized in fig. 11.

Some of the other physiological functions of corticoid hormones will be discussed further in chapter 10, in which we shall discuss the significance of the adrenal glands in adaptation to the environment.

The cellular components on which cortisone acts
In order to understand the way in which some hormones can exert their effects by altering the genetic material of the cell we must first consider the way in which genes influence cell activities.

The genetic material of the cells residing in the nucleus controls cell activities by determining the kind and amount of those protein molecules which regulate the chemical processes of the cell—that is the enzymes. Enzymes are composed of mixtures of the twenty different amino acids, linked together in a specific sequence for each kind of enzyme. The enzymes are synthesized on those thousands of small particles contained by every living cell which are called ribosomes. Information is supplied to the ribosomes for the manufacture of particular enzymes in the form of a long molecule of ribose nucleic acid—messenger RNA. The message of messenger RNA is written in a four-letter code, that is in terms of the sequence of arrangement of the four RNA monomers (uracil, cytosine, guanine, adenine) along the molecule. Ribosomes decode this message and translate it in terms of synthesis of protein.

Messenger RNA is manufactured on the nuclear chromatin using the DNA of the chromatin as a template. Every cell of the body contains the same kind and amount of DNA. However, in different types of cells different types of protein, different enzymes, are produced. In a particular cell the enzymes produced thus form only a fraction of the total number of enzymes which the cell is capable of producing. The nucleus contains the processes which determine which genes are active in the production of appropriate messenger RNA. This is the basis of cellular differentiation. The cells of the exocrine secreting cells of the pancreas produce the specific enzymes of pancreatic secretion, the activities of the erythrocytes are directed almost exclusively to the synthesis of the protein haemoglobin, and muscle cells produce large amounts of the proteins actin and myosin which form the basis of the contractile myofilaments. All body cells contain the appropriate genetic information for the manufacture of these proteins, but in most body cells this information is not put to use, i.e. the particular genes are repressed.

Further, even in a particular differentiated cell the activity of the genetic apparatus is variable depending upon various factors, including the quantity of particular substrates available. This variation in time of gene activity is the basis of cellular adaptation, and the influence of hormones on this plastic genetic material provides the basis of adaptation of the whole organism to changing environmental influences. Repressed areas of chromatin are covered by a layer of those basic proteins called histones of which there are several types in each cell. In some tissues the act of derepression of the genetic material is associated with a loss of histone from the chromatin. The genes are, as it were, functionally and morphologically uncovered.

The evidence that cortisone acts on the chromatin of the cell

For many years it has been known that following the injection of cortisone into various mammals there is a marked increase in the amounts of several enzymes in the liver cells. These enzymes include glutamic-tyrosine transaminase, glutamic-alanine transaminase, tryptophan pyrolase, aldolase, fructose diphosphate phosphatase and glucose-6-phosphatase. In 1962 Kenney treated rats with cortisone and also with radioactively labelled amino acids. He later killed the rats and isolated the enzyme glutamic-tyrosine transaminase from the liver. He found that the enzyme he isolated was also radioactive, indicating that it had been synthesized from administered amino acid under the influence of cortisone.

The effect of cortisone on liver enzymes can be blocked if animals are also given the substance puromycin. Puromycin is an amino acylnucleoside which partially mimics the action of that cytoplasmic RNA known as transfer RNA which is responsible for gathering up appropriate amino acids from the cell and bringing them into close contact with the surface of the ribosomes where they are synthesized into protein under instruction from the molecules of messenger RNA. Puromycin prevents the normal growth of the peptide chain on the ribosome and thus specifically inhibits protein synthesis. Clearly, then, the increased amounts of enzymes in liver cells after treatment with cortisone is due to the synthesis of new protein on the ribosomes.

In addition to producing an increase of amounts of various enzymes in liver cells cortisone treatment also results in an early increase in the rate of nuclear RNA synthesis which is followed by the appearance of increased amounts of cytoplasmic RNA, including messenger RNA. If one prevents this effect of cortisone, then one also prevents the increased formation of enzymes. The effect of cortisone on RNA synthesis and thus on enzyme synthesis can be blocked by the administration of the antibiotic actinomycin D (fig. 12). Actinomycin D base pairs specifically with guanine as it is found in the double helix of DNA and the complex so formed prevents the enzyme RNA polymerase from using DNA as a template for RNA synthesis.

Although cortisone undoubtedly has the above-mentioned effect on the synthesis of liver enzymes the significance of this effect for the regulation of blood glucose is still a matter of debate. Fig. 13 shows

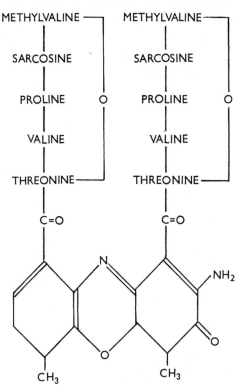

FIGURE 12. Structure of actinomycin D.

the results of an experiment using rats to test the effect of actinomycin
on the changes in liver glycogen and blood glucose which are produced by treatment with hydrocortisone. Four groups of rats were
taken and food was withheld from them for 12 hours.

Group 1. Starved control rats. No treatment except a control
injection of saline.
Group 2. Starved and treated with actinomycin D.
Group 3. Starved and treated with actinomycin D and hydro-
cortisone.
Group 4. Starved and treated with hydrocortisone.

The bar graphs show the glucose concentration in blood (mg %),
liver glycogen content (G %) and the activity of a liver enzyme–
phosphoenolpyruvate carboxykinase (PEP) in each group of animals.
The enzyme PEP is only one of many which are involved in gluconeogenesis, but it was chosen for study because it is a key catalyst in

gluconeogenesis. The results show that in the normal animal fasting produces an increase of PEP from 50 units in the normal-fed animal to 90 units after fasting for 12 hours. Treatment with actinomycin prevented the response of PEP *but* did not prevent the animal from maintaining a normal glucose concentration in blood. Treatment with hydrocortisone, as expected, increased the amounts of glucose in blood and glycogen in liver, and increased the activity of PEP. The addition of actinomycin to hydrocortisone prevented the increase of

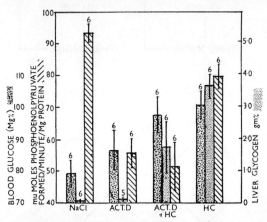

FIGURE 13. The effect of hydrocortisone on hepatic glycogen, blood glucose and hepatic phosphoenolpyruvate carboxykinase in rats treated with actinomycin D prior to fasting. Standard deviations are represented by the vertical lines. The number of rats contributing to each value is given above the standard deviation.

Included by courtesy of Professor P. D. Ray from P. D. Ray *et al.*, 1964. Mode of action of Glucocorticoids. 1. Stimulation of gluconeogenesis independent of synthesis *de novo* of enzymes. *Journ. Biol. Chem.*, 239, 3396–3400.

PEP activity above the normal level (50) but did not significantly reduce the influence of hydrocortisone on blood glucose, although it did reduce the response of liver glycogen.

These results show clearly that in short-term experiments the ability of hydrocortisone to increase the glucose content of blood in fasting is not significantly reduced if the action of the hormone on RNA synthesis (and thus on enzyme synthesis) is blocked by the agent actinomycin. In experiments repeated on rats from which the adrenal glands had been removed similar results were obtained except that in this situation the effects of hydrocortisone on liver glycogen were reduced by 60% by the administration of actinomycin.

One must conclude that cortical hormones must have other actions to promote gluconeogenesis in addition to their effects on messenger RNA production in the liver. One such action is to increase the availability of amino acids to the enzyme systems in the liver—that is to increase the supply of raw material for glucose production. Cortical hormones affect protein metabolism in many body tissues, reducing the rate at which amino acids are incorporated into protein. These amino acids now become available for gluconeogenesis.

Summary

Thus two mechanisms of action of adrenocortical hormones at the molecular level must be considered (Weber *et al.*, 1965):

1. The hormones release gluconeogenic precursors (e.g. amino acids) from peripheral tissues which become available for glucose synthesis.
2. The hormones stimulate the hepatic synthesis of enzymes concerned with gluconeogenesis. This effect does not seem to start immediately and the apparent effects on gluconeogenesis are somewhat delayed.

According to our present understanding of the process both mechanisms appear to be involved although the primary effect would appear to be on the peripheral tissues. However, it seems safe to say that induction of liver enzymes is an integral part of a maximal gluconeogenic response to adrenal corticoids (Ray, 1968).

Other mammalian hormones acting at the gene level

A variety of other mammalian hormones are known to exert their effects at the level of the gene. The various lines of evidence which support this view include the following:

1. An early increase in the rate of nuclear RNA synthesis following administration of the hormone.
2. The appearance of increased amounts of cytoplasmic RNA, in particular that with the characteristics of messenger RNA.
3. Inhibition of the effects of the hormone by the agents actinomycin D and puromycin.
4. The appearance of new types of cytoplasmic protein following administration of the hormone.

Evidence is available that the effects of insulin on protein metabolism, some effects of growth hormone, thyroxine, oestrogens, testosterone, pituitary ACTH and aldosterone are mediated by way of the chromatin of the cell.

It is important to recognize that although several hormones may produce effects at the same site in cells they may do so by quite different mechanisms. Thus both aldosterone and vasopressin increase the rate of transport of sodium and water across the cells of the toad bladder, but do so by different mechanisms. The effect of aldosterone but not that of vasopressin is blocked by actinomycin.

The action of hormones at the level of the gene.—II. Ecdysone

The hormone ecdysone is considered here because not only does it produce changes in the activity of genes but also morphological changes in chromatin which can be readily observed with the light microscope.

Ecdysone is the 'moulting and metamorphosis' hormone of insects. The development of insects is punctuated by a series of moults which may be larva–larva, larva–pupa or pupa–imago. These moults involve the co-operation of several endocrine tissues, neurosecretory cells of the brain, the prothoracic glands and corpora allata. Ecdysone is produced by the prothoracic glands after this gland has been activated by a hormone from the brain. Ecdysone regulates the events of the moult. Another hormone called juvenile hormone determines the kind of moult initiated by ecdysone. When juvenile hormone is present larval characters are maintained and the moult is larva–larva. In the absence of juvenile hormone the moult will be either larva–pupa or pupa–imago.

The mode of action of ecdysone has been studied in Dipteran larvae. These species have a cytological peculiarity which has permitted the locus of action of ecdysone to be demonstrated in a direct fashion. This cytological peculiarity is the presence of giant or polytene chromosomes in salivary gland and other tissues (fig. 14). When fully grown these chromosomes are $100 \times$ thicker and $10 \times$ longer than the chromosomes of other body tissues. Polytene chromosomes are present in tissues which grow by means of an increase in cell size rather than cell number. As the cells of these tissues enlarge the chromosomes undergo repeated longitudinal replication and also increase in length. Each chromosome shows a characteristic pattern of transverse bands which contains large amounts of DNA and histone. Each band is a gene or a small group of genes.

During the development of a larva certain bands on the chromosomes become swollen while others remain unswollen. These swellings, called puffs, are areas of intense RNA synthesis. Actinomycin D which blocks DNA dependent RNA synthesis will prevent this puffing. We can regard puffed areas of the chromosomes as derepressed genes. During the course of moulting there is a characteristic sequence

of puffing of different areas of the chromosomes. The same sequence of events follows the injection of a minute quantity of ecdysone into a larva which is not preparing to moult. It is possible to induce puffing in a Dipteran larva by the injection of only $2 \cdot 1^{-6}$ mg ecdysone and the first signs of puffing appear after only 15 minutes. This effect of ecdysone is blocked by actinomycin D but not by puromycin. Using very elegant techniques it has been possible to analyse the purine base composition of the RNA produced by different puffs; the base composition varies from puff to puff as one would expect since they are the products of different genes. It is thus clearly established that moults are regulated by ecdysone which produces its effects by derepressing various genes. Gene activation expresses itself in terms of messenger RNA production leading to protein synthesis by the ribosomes and the establishment of the metabolic changes associated with moulting.

The appearance of a new kind of messenger RNA following treatment with ecdysone was shown in a very elegant fashion by Sekaris and Lang in 1964. In the last larval moult which produces the pupa of the blow-fly the pupal skin hardens and darkens. A hardening agent, N-acetyldihydroxyphen-ethylamine appears in the epidermis. The hardening agent is derived by decarboxylation of dihydroxyphenylalanine (DOPA) under the influence of the enzyme DOPA decarboxylase. Sekaris and Lang isolated the nuclei from epidermal cells after larvae had been treated with ecdysone. RNA was then extracted from the epidermal cell nuclei and this was added to an isolated system of mammalian ribosomes in which all the factors needed for protein synthesis were present with the exception of messenger RNA. The addition of RNA from blow-fly larval epidermis to this system resulted in the synthesis of DOPA decarboxylase by the ribosomal system. Thus the larval RNA has the features of a particular messenger RNA. RNA from control larvae not treated with ecdysone produced no detectable DOPA decarboxylase by the ribosome system.

It has been suggested that ecdysone produces its effects on the genes by altering the permeability and transport functions of the cell membrane, or the nuclear membrane, or both (Kroeger, 1967). If a microelectrode is inserted into a chironomid salivary gland cell one can record an electrical potential across the cell membrane. This is presumably due to a gradient of ion distribution between the cell interior and surrounding tissue fluid. If ecdysone is added to the incubation medium containing isolated salivary gland this causes an increase in the recorded trans-membrane electrical potential. This change occurs within one minute of the addition of the hormone. This effect of ecdysone is presumably due to a change in the distribution of charged ions. Striking confirmatory evidence to support this view of the mode of action of ecdysone is provided by the observation that the effect of ecdysone can be mimicked by altering the ionic environment of salivary gland cells. Some chromosome puffs appear in salivary gland cells after an increase in the extracellular concentration of one of several ions (e.g. Na^+, Mg^{++}, Ca^{++}, NH^+_4). Other puffs appear only following increases in the concentration of one ion. In view of the rapidity of the effects of ions and because they are as

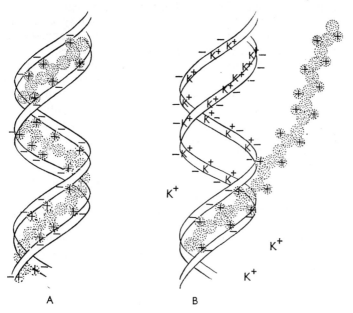

A   B

FIGURE 15. Representation of a suggested mode of action of ecdysone. In A the groove in the double helix of DNA is occluded by the presence of nucleo-histone, the positively charged side chains of which are bound to the negatively charged phosphate groups in the backbone of the DNA molecule. In B there is an influx of positively charged ions into the nucleus under the influence of ecdysone which compete for and displace nucleohistone from stretches of DNA. This denuded DNA can now act as a template for the synthesis of RNA.

effective on isolated nuclei as on whole cells (i.e. no cytoplasmic factor is necessary for their action) it has been suggested that the ions may act directly upon the chromatin, possible upon DNA-histone links. Thus ions could compete with histones for the negative charges on the DNA molecule and cause a displacement of histone from its position on DNA. This would result in the revealing of stretches of DNA which could now act as a template for the synthesis of RNA (fig. 15).

Growth hormone: an action on ribosomal translation of messenger RNA

The discovery of messenger RNA and the publication of Jacob and Monod's concepts of the regulation of gene activity in bacteria were followed by enthusiastic application of these ideas to the problems of mechanisms of hormone action. We have seen how evidence has been obtained that some effects of hormones on protein synthesis are associated with the derepression of genes and the synthesis of new kinds of messenger RNA. In the larvae of Lepidoptera these changes can be visualized on the giant polytene chromosomes.

It is, however, becoming clear that primary action at the gene level is by no means the only way in which hormones can modify the protein synthesis of the cell. Indeed, there is evidence that messenger RNA is not necessarily a limiting factor in protein synthesis. In the egg of the sea urchin the act of fertilization results in a rapid increase in protein synthesis. But this occurs in spite of a complete absence of RNA synthesis. Messenger RNA must have been present in the egg before fertilization, although this was not associated with protein synthesis. Other studies of protein synthesis in a liver tumour showed that the rate of synthesis was very similar to that of normal liver tissue in spite of the fact that the tumour contained over four times as much RNA as normal liver. We shall consider here some of the evidence that pituitary growth hormone exerts at least part of its effects on protein synthesis by modifying the ability of ribosomes to translate messenger RNA into terms of protein synthesis.

The pituitary gland and growth

No attempt will be made here to give a complete account of growth hormone, its physiological effects, mode of action and the regulation of its secretion from the anterior pituitary gland. The physiological consequences of administering the hormone to animals are complex and involve not only widespread effects on the metabolism of protein, carbohydrates and lipids but also interrelations with other hormones such as insulin and glucocorticoids. Accurate estimations of the quantity of the hormone in body fluids has only recently become possible. The true physiological significance of growth hormone is for these various reasons still a matter of controversy and speculation. It has indeed been doubted that the hormone has a direct effect upon growth, and it has been suggested that the growth-promoting effects of the hormone are due to its ability to stimulate the secretion of another hormone—insulin.

We can define growth as an increase in body weight and size due to the laying down of tissues which have a similar composition to those of the original body. This definition thus excludes changes in body weight and size due to accumulation of water or fat. More precise indices of growth of an animal are increases in the amount of body protein or the size of bones. The significance of the pituitary gland for growth was first indicated by observations of a rare disease of man first described in 1886 in which 'overgrowth' of various tissues (bones, skin, tongue, abdominal organs) of the adult are associated with a particular tumour of the anterior pituitary gland.

This condition is called acromegaly. In acromegaly there is little extra growth of long bones (femur, tibia, etc.) because these bones grow in length only at the disc of cartilage which separates the shaft from the head of the bone. This cartilaginous disc is transformed into bone during adolescence. If overactivity of the anterior lobe of the pituitary gland occurs before this transformation has occurred, then marked growth of long bones becomes possible with the production of a giant. The reverse condition of dwarfism results from deficiencies of pituitary growth hormone.

The pituitary gland influences growth in various ways. First, by regulating other endocrine glands—ovary, testis, thyroid—the secretions of which influence growth. Second, by the production of a specific growth hormone, also called somatotrophin.

Some effects of administration of growth hormone

In animals in which the epiphyses of the long bones are not united with the shaft the administration of growth hormone stimulates the growth of both bone and soft tissues to produce a symmetrical enlargement of the animal. Treatment of rats with growth hormone can produce individuals which are twice the size of normal individuals, such is the potency of the hormone. Nearly all organs and tissues participate in the response to growth hormone. Biochemical studies show that this growth is due to an increased synthesis of protein from amino acids. At the cellular level growth hormone stimulates the synthesis of RNA of all types, nuclear RNA, transfer RNA, messenger RNA and ribosomal RNA.

The effects of removal of the anterior pituitary gland are the reverse of the above changes. In the growing animal there is a virtual cessation of growth. In adults there is a loss of body weight. Tissues removed from these animals show depression of the incorporation amino acids into protein, a fall in the RNA content of cells and a decline in the number of ribosomes. The administration of purified growth hormone can reverse these changes.

The mechanism of action of growth hormone on protein synthesis

The biochemical effects of growth hormone, as described above, would be consistent with the idea that growth hormone acts at the genetic level by promoting increased synthesis of messenger RNA, ultimately leading to increases in protein synthesis. The rate of synthesis of RNA can be studied in isolated tissue preparations by studying the rate of incorporation of radioactive RNA precursors into RNA. These studies showed that the administration of growth hormone stimulated the labelling of RNA. These effects are similar to those produced by hormones such as ecdysone, oestradiol and testosterone.

The effects of removal of the anterior pituitary gland provided further support for an action of the hormone on chromatin. This procedure in the rat results in a decrease in the number of polysomes (aggregations of ribosomes and messenger RNA) in the liver which suggests that removal of growth hormone had reduced RNA production and hence formation of polysomes.

This view of the action of growth hormone was dealt a blow by the discovery that although the effect of growth hormone on the synthesis of messenger RNA could be blocked by actinomycin this did not prevent the stimulation of protein synthesis by ribosomes of liver or diaphragm. We have already seen that the amount of messenger RNA in cells is not necessarily a limiting factor in protein synthesis. Further support for this is provided by experiments studying the effects of removal of the anterior pituitary gland on the rat heart. This procedure reduced the rate of protein synthesis of heart muscle although the number of polysomes was not altered.

It would seem that growth hormone can stimulate protein synthesis in a way which does not necessarily involve an increase in the rate of production of messenger RNA. Korner has provided evidence that a further site of action of growth hormone in the cell is the ribosome. The removal of the pituitary gland reduces the ability of ribosomes to assemble amino acids into protein and this effect can be reversed to a degree by treatment with growth hormone. This change in the activity of the ribosomes seems to be associated with a reduced ability to attach themselves to messenger RNA.

We can conclude that the action of growth hormone on protein synthesis is not readily explainable by 'switching on' of particular genes. Certainly increases in RNA synthesis occur after treatment with growth hormone, but this may well be secondary to actions of the hormone in the cytoplasm, in particular to an increased association of ribosomes and messenger RNA. Obviously further information is needed to unravel this problem of growth hormone action.

4

Local Hormones: Progesterone and the Defence Mechanism of Pregnancy

----

Examples of local hormones have been mentioned in various sections of the book, e.g. the activation of adrenal medullary cells by the release of acetylcholine at nerve terminals (p. 104) or the regulation of anterior pituitary function by means of neurohumors liberated into the hypothalamico-hypophyseal portal circulation (p. 143). However, although classical local hormones such as histamine, 5-hydroxytryptamine, bradykinin, substance P and prostaglandins have been known for many years our understanding of their significance in the organism is very incomplete. One general hormone, progesterone, also appears to act as a local hormone in the uterus where it has considerable significance in the maintenance of pregnancy. This local effect of progesterone will be the subject of this chapter.

Sources of progesterone

The steroid hormone progesterone is produced by several tissues of the body and is concerned in various aspects of reproductive function. In the normal oestrous or menstrual cycle, uninterrupted by pregnancy, progesterone is produced by the corpora lutea of the ovary—those temporary endocrine glands which develop from the Graafian follicles after they have ruptured and liberated the ova. The functional life of these endocrine glands varies from species to species, but it is short unless pregnancy supervenes. The most significant function of progesterone produced by the corpora lutea is to promote, together with oestrogen, those changes in the mucous membrane lining the uterus which facilitate the implantation of the fertilized ova. If fertilization and implantation occur, then the corpora lutea continue to secrete progesterone. This is because the placenta, even early in its development, produces large amounts of a pituitary-like gonadotrophin (chorionic gonadotrophin) which supports the continued development and secretory activity of the corpora lutea. Later in the pregnancy the placenta itself contributes directly by the secretion of considerable amounts of progesterone.

Functions of progesterone

Progesterone is concerned not only with the preparation of the endometrium for the implantation of fertilized ova but it is also necessary for the continued maintenance of pregnancy. As early as 1928 Parkes showed that corpora lutea are necessary for the maintenance of pregnancy in mice. Parkes prepared mice for the experiments by destroying one ovary early in life by exposure to X-rays. When the mice were sexually mature they were mated, became pregnant, and then between the 11th and 17th day of pregnancy the healthy ovary was removed from a group of mice. This was followed by premature delivery of the foetuses. However, he also found that the removal of the sterilized ovary, containing no corpora lutea, did not disturb the pregnancy. These findings were later confirmed in rats, rabbits and other species. Extracts of corpora lutea or pure progesterone can prevent the termination of pregnancy which follows the removal of the ovaries.

Treatment with progesterone not only can replace the action of the ovaries in the maintenance of early pregnancy but can, if injected in adequate amounts, prolong the pregnancy beyond its normal duration. In 1934 Portman treated rabbits with 2–3 units of progesterone towards the end of gestation and was able to prolong the duration of pregnancy by 2–3 days, the young being born alive. With daily doses of 4 units, gestation was prolonged by up to 9 days although the young were born dead. These findings have been repeatedly confirmed, including modern studies using more potent progestational compounds. In animals in which pregnancy is prolonged by the administration of progesterone the ovaries may show degenerate corpora lutea, showing that these could not have been the cause of the postponement of labour.

Hormones and the activity of the myometrium

The target organ which ultimately determines the maintenance of pregnancy and its effective termination at the appropriate time is the myometrium—the mass of smooth muscle cells of the uterus. During the oestrous cycle and pregnancy there are pronounced changes in the activity of the myometrium and its working capacity.

During the oestrous cycle itself the uterus shows greatest activity during the actual period of oestrus, when the endocrine picture is dominated by oestrogen secreted by the Graafian follicles. If the uterus is removed from an animal in the state of oestrus and is placed in a bath of warm physiological saline solution and connected to a recording device, the uterus is seen to show rhythmic contractions of maximal amplitude. This activity is independent of any nervous or pharmacological stimuli. The uterine activity can also be observed in the intact animal, e.g. by means of a transparent plate sewn into the body wall. At other times of the oestrous cycle the activity of the uterus is irregular and submaximal in amplitude. When pregnancy develops the uterus becomes mechanically inert and remains relatively quiescent until the onset of labour. During pregnancy, in spite of the quiescence of the uterus its working power increases and during labour powerful rhythmical contractions of the myometrium effectively discharge the foetuses and placentae.

In addition to these changes in spontaneous activity of the uterus there are also marked fluxes in the ability of drugs and physiological stimuli to initiate myometrial activity. The sensitivity of the uterus to the physiological stimulant oxytocin is maximal in oestrus and in early labour. Thus in the pregnant rabbit 48 hours before the onset of labour the activity of the uterus is slight and intra-venous injection of 25–50 milli-units of oxytocin is needed to produce a noticeable increase in myomterial activity. During the next 48 hours the sensitivity of the uterus to oxytocin increases markedly so that the intra-venous injection of 0·5–1·0 milli-units produces observable increases in uterine activity. These observations were made by Anna-Riitta Fuchs (1966) by means of water-filled balloons inserted into the cavity of the uterus and connected by a nylon catheter to a pressure recorder.

The myometrial cell

The target for progesterone acting either as a general hormone in the circulating blood or acting locally on the uterus from the placental site is the smooth muscle cells of the uterus. Before considering the possible local action of progesterone in the maintenance of pregnancy it is necessary to consider something of the nature of the target organ.

The smooth muscle cells of the uterus resemble in many ways smooth muscle in other parts of the body but differ in some special respects. For descriptive purposes we can regard the smooth muscle cell as a bag of excitable membrane enclosing the contractile machinery. Across this excitable membrane we can record a potential gradient, the inside of the cell being electrochemically negative to the extracellular fluid. This trans-membrane potential is generated by the active extrusion of sodium ions from the interior of the cell, and because of the low permeability of the resting membrane to sodium ions these do not regain access to the interior of the cell as fast as they are extruded. The membrane potential of visceral smooth muscle is characteristically low (30–60 millivolts) compared to that of striated muscle (90 millivolts). This difference is largely responsible for one striking difference in the activity of smooth and striated muscles, the presence or absence of spontaneous mechanical activity.

Excitation of muscle cells is, in mammals, produced by a lowering of the trans-membrane potential to a critical level at which an explosive event occurs in the cell membrane during which its properties change radically. The membrane suddenly becomes markedly permeable to the sodium ion, and these ions move into the interior of the cell along an electrochemical and concentration gradient; the resting potential is abolished and is even reversed, the interior of the cell becoming electropositive to the exterior. This change is propagated along the excitable membrane and is recorded electrically as the action potential. Associated with these changes in the electrical activity of the membrane there is an influx of calcium ions into the interior of the cell which activate the contractile machinery.

Striated muscle cells have a high stable resting membrane potential. They are inactive unless nerve stimulation releases acetylcholine at the motor end plates. The acetylcholine produces an increased permeability of the membrane to all ionic types and these move along their concentration gradients and short circuit the membrane. This process depolarizes the end plate and when this reaches a critical level then the explosive event occurs and a wave of depolarization sweeps along the entire muscle cell membrane. Striated muscle cells can thus be regulated in one direction only, i.e. by depolarization.

The membrane potential of smooth muscle is lower than that of striated muscle and is closer to the level of membrane potential at which action potentials are generated. Moreover, individual smooth muscle cells may show spontaneous shifts of membrane potential which bring the membrane into the level of 'firing' of action potentials (fig. 16). Smooth muscle cells, unlike striated muscle cells, are functionally connected to one another so that the development of spontaneous activity in one cell may propagate to adjacent cells resulting in contraction of a whole stretch of muscle. These features are responsible for the spontaneous activity of visceral smooth muscle, that is activity in the absence of any type of external stimulation. Smooth muscle is innervated although the role of nervous activity

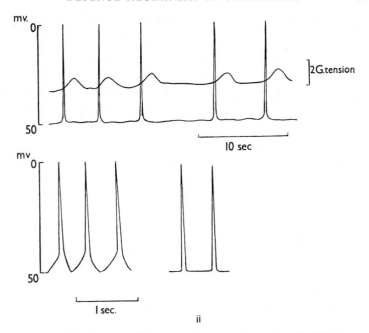

FIGURE 16. Action potentials of uterine smooth muscle cells. (A) Intra-cellular recording of action potentials and mechanical activity in a 20-day pregnant rat uterus. The upper trace shows contractions which follow the discharge of action potentials. (B) Action potentials recorded by a microelectrode inserted into a smooth muscle cell of the uterus; (i) shows a slow depolarization before generation of the action potential. This is a record from a 'pace-maker' cell which generates spontaneous activity. During the slow depolarization the membrane becomes in-creasingly permeable to sodium ions; these enter the cell and cause a fall in the resting potential and when this reaches a critical value an action potential is generated. This electrical activity may be conducted to adjacent cells via low resistance bridges between the cells; (ii) shows action potentials in these conducting cells which do not show the pre-potential.

is the integration and modulation of the spontaneous activity. One further difference between smooth and striated muscle cells lies in the number of biological and pharmacological activators which can stimulate the muscle. Striated muscle is activated by means of acetylcholine acting at discrete sites on the muscle membrane, the motor end plates. Smooth muscle may be activated by a whole range of substances; uterine smooth muscle for example can be stimulated

c

by acetylcholine, adrenaline, noradrenaline, histamine, bradykinin, 5-hydroxytryptamine, vasopressin, oxytocin, etc. Further, the activity of smooth muscle can be modulated in two directions; the membrane potential poised as it is at an intermediate level can be reduced to the critical level at which excitation occurs or it can be raised above this critical level so producing relaxation of the muscle. Thus the administration of adrenaline or noradrenaline can relax uterine muscle and this effect is associated with an increase in the resting potential. Uterine smooth muscle shows additional features in that the sex hormones, oestrogen and progesterone, can change the resting potential and the tendency to generate action potentials. Further, these hormones can alter the working capacity of the myometrium by producing an increased production of the contractile proteins.

If a microelectrode is inserted into a uterine smooth muscle cell in the immature rat the resting transmembrane potential is found to be low, the interior of the cell being 45 millivolts negative to the exterior. These uteri are inactive. After oestrogen has been injected into the animal the membrane potential increases to about 57 millivolts. At this level of membrane potential spontaneous generation and conduction of action potentials occurs and spontaneous activity is vigorous. The size and frequency of the contractions depends on the frequency of generation of action potentials and by the total number of cells that are active (fig. 16). If progesterone is now administered to the animal further changes occur. The resting potential increases to a value of 63 millivolts. Synchronous discharges of action potentials do not occur and action potentials which do arise are not conducted over long stretches of the myometrium. Associated with these effects the spontaneous activity of the uterus is weak and irregular and there is a depression of the responses to stimulants such as oxytocin. It is these effects mediated by circulating and local progesterone which determine the progesterone defence mechanism of pregnancy.

Placental progesterone and the defence mechanism of pregnancy

Various workers have provided evidence that during pregnancy the effects of progesterone circulating in the blood are reinforced by progesterone diffusing from the placental site into the overlying myometrium. This local effect of placental progesterone produces in the adjacent myometrial tissue an exaggeration of those features shown by the remaining interplacental parts of the myometrium.

In the rabbit intracellular measurements of the resting potential of myometrial cells overlying the placenta (20–26th day of pregnancy) gave a mean value of 53 millivolts, whereas at interplacental sites the value was 42 millivolts. This difference in membrane potential between placental and interplacental parts of the myometrium can be abolished if sufficient amounts of progesterone are administered to the animal.

Differences in the activity of the various parts of the pregnant myometrium have also been demonstrated, by means of experiments in which the foetuses were removed, leaving the placentae intact, the foetuses being replaced by wax dummies. The wax dummies were subsequently removed and found to be narrower at points of contact with the interplacental portions of the myometrium while remaining unchanged at the placental portion. These effects can be explained by a greater mechanical activity of the interplacental parts of the myometrium.

If the placenta is indeed concerned in the production of areas of myometrium which are mechanically inactive, insensitive to circulating stimulants, and which prevent the conduction of excitation throughout the myometrium, then disturbances of placental function should be followed by changes in the activity of the myometrium. Placental function can be depressed by the injection of hypertonic solutions of saline into the amniotic cavity. This procedure results in the premature onset of labour. If the disturbance of placental function is carried out in one uterine horn only, then only this horn empties its contents.

The onset of labour

The onset of labour, that is the change from a quiescent uterus to one showing rhythmical propagated activity, does not seem to be initiated by a sudden withdrawal of the progesterone defence mechanism. In those species in which the chief source of progesterone at the end of pregnancy is the placenta there are no detectable changes in the progesterone content of the circulating blood prior to the onset of labour.

It appears that the progesterone block of myometrial activity provides only a relative defence. The concentration of progesterone in the myometrial cell is regarded as a function of the distance of the cell from the placenta. The cells nearest to the placenta are those most influenced by the diffusion of progesterone from the placental bed. However, during pregnancy, as the uterus enlarges, the ratio of nonplacental: placental parts of the myometrium increases. Thus as pregnancy advances, increasing areas of the myometrium are removed from the direct influence of placental progesterone. These areas become increasingly active and sensitive to circulating stimulants. When the active area reaches a critical size, propagated electrical and mechanical activity occurs and the pregnancy is terminated. However, even at this late stage, the administration of progesterone, particularly if injected directly into the myometrium, can temporarily arrest myometrical contractions.

We can conclude that there is appreciable evidence to implicate local progesterone from the placenta in the regulation of uterine activity during pregnancy and in determining the onset of labour. The onset of labour may be reinforced by the liberation into the circulating blood of the uterine-stimulating hormone oxytocin from the posterior pituitary gland. The precise rôle of oxytocin in determining the onset of labour is, however, subject to some controversy and this is not the place to discuss the conflicting experimental observations.

5

Hormones and Homeostasis— I. Regulation of Calcium Metabolism

---

Hormones and homeostatic functions

The term homeostasis was coined by Cannon in the 1920s to describe the function of many physiological processes directed at stabilizing the internal environment of the body. The importance of stability in the internal environment was stressed by Claude Bernard (1879). It was Bernard who recognized that there are in fact two environments for an animal, an external environment and a milieu interieur (an internal environment in which the tissues live). He put forward the view that 'the invariability of the internal environment is the essential condition of full independent life'. Stability of the internal environment, he said, implies a perfection of the organism such that external variations shall be, at every instant, compensated and brought into equilibrium. In the mammal, highly complex mechanisms exist which stabilize such features as temperature, blood pressure, the partial pressure of oxygen and carbon dioxide in the blood, and the concentration of hydrogen ions, sodium, potassium, calcium and glucose in tissue fluids. These mechanisms include, among other regulating mechanisms, a variety of endocrine systems.

Body tissues are not equally sensitive to all the changes which occur in the nature of the internal environment. Nerve cells, for example, are particularly sensitive to reduction of oxygen or glucose supplies or to changes in the potassium and calcium ions in the fluid that bathes them. Tolerance of fluctuations in solute composition of body fluids also varies from one solute to another. The pH of blood plasma is

61

maintained by physiological mechanisms at about 7·4. However, the limits of change compatible with life range from 7·0 to 7·8. This is a wide range of tolerance for changes in [H$^+$] since the pH scale is a logarithmic one, the [H$^+$] at pH 7·0 being 250% of normal, while at pH 7·8 it is only 40% of normal. Toleration of the variation in pH is much greater than the tolerance of variation in the concentration of sodium, potassium or calcium ions. The normal concentration of plasma calcium is 10 mg %. If the concentration falls to 5–7 mg %, then spontaneous twitching of muscles occurs, which progresses to generalized convulsions and to death by respiratory failure. The normal concentration potassium in plasma is 4·5 milli-equivalents per litre and if this rises to 8 milli-equivalents per litre, there is increasing danger of death from ventricular fibrillation, a condition in which the contractions of the ventricle become unco-ordinated. Thus the mammal will tolerate changes in [H$^+$] amounting to about 200% of normal, but will not tolerate changes in calcium exceeding about 30%, nor changes in potassium exceeding about 50%. Although the mammal will tolerate wide changes in [H$^+$] when this is drastically altered during experiments, under normal conditions the control of [H$^+$] is precise and small changes in [H$^+$] of plasma result in physiological compensations to stabilize [H$^+$]. In fact, the mammal is said to be a million times more sensitive to absolute change in [H$^+$] than to absolute change in [Na$^+$].

The endocrine regulation of calcium metabolism

Calcium content of the body

The body of an adult man contains about 1 kg of calcium. About 99% of this is present in the skeleton. The small amount of calcium in the blood (concentration 10 mg%) is present in two forms, an ionized form and a non-ionized form. The ionized form is physiologically active, whereas the non-ionized form which is bound to plasma protein is physiologically inactive. Of the calcium which is ingested, 60–80% is absorbed from the intestine and gains access to the body fluids, where it may become bound to plasma protein, or form other complexes, or it may remain in the physiologically active ionized form and be available for bone formation. Calcium in excess of requirements is excreted in the urine and faeces.

The rôle of calcium in the body

In addition to the obvious mechanical function of calcium in the form of hydroxyapatite, as a component of the skeleton, calcium ions are involved in many vital processes at both cellular and extracellular levels. In blood, calcium is indispensable in the coagulation process in which a plasma protein, fibrinogen, is converted into the insoluble strands of fibrin of the blood clot. At the level of the cell membrane, calcium plays an important part in regulating the permeability characteristics of the membrane. Here it acts as a stabilizer of the membrane, reducing its permeability to ions so that the cell is able to establish concentration gradients or electrochemical gradients across the membrane by actively extruding ions from the interior of the cell. This function is of particular importance to those cells which at rest develop electrochemical gradients across the cell membrane which undergo rapid changes when the cell is excited. When supplies of calcium in the extracellular fluid are inadequate these cells become 'leaky', that is they show an increased permeability to ions, which can now cross the cell membrane along their concentration or electrochemical gradients. This results in a fall of the normal electrochemical potential between the interior and the exterior of the cell.

Voluntary muscle fibres at rest show a high stable resting transmembrane potential of 90 millivolts. The interior of the cell is electrically negative to the exterior surface because of a net deficiency of intracellular cations (sodium), which are being continually and actively extruded from the cell. Because of the effect of calcium ions on the cell membrane these extruded ions are not able to diffuse back into the cell, or at least they cannot diffuse back in as fast as they are being extruded, and so they permit the establishment of the resting potential. In conditions of calcium deficiency, the resting potential falls and the muscle cell becomes spontaneously active as the resting potential falls to the critical level at which normally an action potential is generated in a cell membrane (see p. 56). This involuntary spontaneous twitching is known as tetany. A deficiency of calcium ions affects nerve cells in a similar way to voluntary muscle fibres, causing an increased excitability.

Calcium ions perform a further function in muscle cells. They form the coupling mechanism which links the electrical events at the cell membrane with the actual contractile elements inside the cell. The generation of an action potential at the cell membrane of the muscle cell results in the liberation of calcium from sites where it has been bound to the membrane. The calcium ions diffuse into the cell and initiate contraction. In resting muscle there is little or no free intracellular calcium, since this is avidly bound by a substance called a relaxing factor. The influx of calcium ions on activation of the cell membrane swamps this factor and permits the calcium to gain access to and to activate the enzyme myosin-ATP-ase. Fission of ATP occurs and sliding of the myofilaments develop.

Activation by calcium ions can occur in cells other than muscle.

There is evidence that in the posterior pituitary and the adrenal medulla, the release of hormones into the circulatory system is calcium dependent.

In view of these rôles of calcium, in particular the regulation of the activity of neurones and muscle, it is not surprising that the concentration of calcium ions in the extracellular fluid is maintained with remarkable precision. In achieving this there is a complex interplay of several hormones acting at several sites.

The parathyroid glands and calcium metabolism

The parathyroid glands, four small yellow bodies applied to the posterior surface of the thyroid glands, were described by several investigators including Ivor Sandström, who gave them their name in 1880. Sandström's observations attracted scant attention from physiologists and the glands were 'rediscovered' in 1891 by Gley who examined the effects of removal of the glands. We now know that there are considerable variations in the distribution of these glands, not only between species but also between individuals of the same species. Accessory parathyroid tissue is known to exist in some individuals and when this is located in the thorax it is easily overlooked. These features account for the early controversies of the importance of the glands.

In 1909 MacCallum and Voegtlin described a fall in the concentration of calcium ions in the blood plasma following parathyroidectomy. The importance of calcium and the symptoms of hypocalcaemia had been previously described so that the relationship between the parathyroids and calcium metabolism was readily established. After several workers had unsuccessfully attempted to obtain parathyroid extracts having physiological activity, Collip in 1925 prepared the first stable acid extract from the parathyroid glands of cattle. The extract when injected into parathyroidectomized animals relieved the hypocalcaemic tetany, and when injected into normal animals it caused a rise in plasma calcium levels. Shortly after this, hypoparathyroidism and hyperparathyroidism (Allbright's disease or Von Recklinghausen's syndrome) were recognized as human diseases.

CHEMISTRY OF PARATHYROID HORMONES

The nature of parathyroid hormone, parathormone was not clearly established until 1960 when Craig and Rasmussen, using various elegant techniques, were able to isolate the pure hormone from phenolic extracts of beef parathyroid glands. They found that the pure hormone was a small protein or polypeptide of molecular weight 9,500, containing seventeen different amino acids in a chain which was 83 amino-acid units long. The exact sequence of the amino-acid residues is not yet established.

THE PHYSIOLOGICAL ACTION OF PARATHORMONE

The main action of parathormone concerns the metabolism of calcium and phosphorus. Before discussing the effects of the hormone, we should stress again that calcium and phosphorus in the organism are present mainly as the very insoluble calcium phosphate and only very small amounts of phosphate or calcium ions can be present in solution in body fluids. The product of the concentration of the ions of calcium and phosphate is a constant and in pure solutions there is a strict reciprocal relationship between them. Thus increase in the concentration of calcium is accompanied by decrease in the concentration of phosphate and *vice versa*. In whole blood the situation is somewhat modified by the presence of other solutes, but in general it is true to say that a change in the concentration of one of these ions is accompanied by an inverse change in the concentration of the other.

The physiological effects of parathormone occur in a distinct sequence of events. The first observed effect of injecting parathormone into an animal is an abrupt and marked increase in the rate of phosphate excretion in the urine, and this may be seen within an hour of administration of the hormone. Following this the concentration of phosphate in the blood plasma begins to fall. This effect is achieved by a reduction in the absorptive capacities of the renal tubules for the phosphate ion. Following the effect on phosphate excretion, the blood calcium begins to rise, reaching a maximum in 12–18 hours after a single dose of parathormone has been given. The urinary excretion of calcium is also increased, but not until much later. Following parathyroidectomy these changes are in the opposite direction, but occur in the same sequence.

There have been two major theories to account for the mechanism by which parathormone causes the release of calcium from bone into the extracellular fluid. Allbright and his colleagues, who made major advances in the understanding of parathyroid pathology and calcium metabolism, believed that the primary effect of the hormone is on the kidney. An increased excretion of phosphate causes a fall in phosphate in the blood and this permits a larger amount of calcium ions to exist in solution, so there is mobilization of calcium from bones.

A second theory stated that the primary effect of parathormone is on the bone, causing its dissolution with a consequent mobilization of calcium ions. One might think that the controversy could readily be resolved by a study of the effect of parathormone on nephrectomized animals, that is after the removal of the kidneys. If parathormone exerted its prime effect on the kidney, then there should be no hypercalcaemia when the hormone is administered following bilateral nephrectomy. However, early experimental work produced conflicting results. For example, Collip and Neufeld in 1942 were unable to produce hypercalcaemia by the administration of parathormone after nephrectomy. However, in 1935 Ellsworth and Frichter had readily obtained hypercalcaemia when they administered parathormone after nephrectomy, and this was obtained even with the development of the hyperphosphataemia which occurs after nephrectomy.

More recent experimental work has clearly established that parathormone has direct effects upon kidney *and* bone, and this work has been carried out with pure hormone. A direct effect on bone was demonstrated by Gaillard using the technique of tissue culture *in vitro*. He cultured thin slices of bone from the skull of a mouse. The slices were cultured in an appropriate medium and then when they were growing, pure parathormone was added. Destructive changes appeared in the bone similar to the lesions seen in human patients who are suffering from disease of the parathyroids in which there is an excess production of parathormone. Similar effects were noticed in other cultures to which living parathyroid glands were added.

A direct effect of the hormone on the kidney was demonstrated by a technique involving the injection of the hormone into *one* renal artery of a dog, while urine was collected separately from each ureter for the estimation of phosphate content. It was found that there was a marked increase in the amount of phosphate excreted by the kidney into which parathormone had been injected, while in the opposite kidney the changes in phosphate excretion were minimal.

At first sight it is a little difficult to make sense of this dual action of parathormone if one regards the primary physiological function of the hormone as a means of increasing the concentration of calcium ions in the plasma when the need arises. One can, however, regard the renal effect of parathormone, that is the increased excretion of phosphate, as a means of reducing the concentration of phosphate in body fluids to offset the liberation of phosphate that accompanies the release of calcium from bone, so permitting a rise in the concentration of calcium ions in the plasma in accordance with the relationship $[Ca] \times [P] = K$.

Recent work with the pure hormone has shown that parathormone acts at two further sites. It increases the absorption of calcium from the intestine, a functional overlap with the 'hormone' vitamin D, and it also increases the amount of calcium reabsorbed from the glomerular filtrate in the kidney. Both of these actions reinforce the action on bone to promote a rise in the concentration of calcium in body fluids. The action of parathormone on the above-mentioned four processes show differences in speed of response and capacity and ensure that the fluids bathing nerve and muscle have a remarkably constant calcium content (fig. 17).

The amount of parathormone liberated from the parathyroid

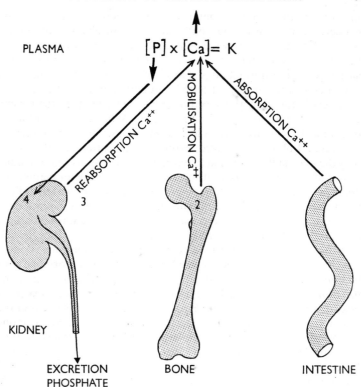

PLASMA

$$[P] \times [Ca] = K$$

REABSORPTION Ca^{++}

MOBILISATION Ca^{++}

ABSORPTION Ca^{++}

KIDNEY

EXCRETION PHOSPHATE   BONE   INTESTINE

FIGURE 17. Summary of the actions of parathormone. Parathormone raises plasma [Ca] by actions 1, 2 and 3. The hormone also reduces plasma [phosphate] by action 4 which offsets the effect of liberation of phosphate which accompanies the release of calcium from bone, so permitting a rise in the concentration of calcium in plasma in accordance with the relationship $[P] \times [Ca] = K$.

glands is determined directly by the concentration of calcium in the blood entering the glands. If the calcium content of the blood falls, the glands respond by increasing the output of parathormone. No humoral or nervous connection with any other tissue is necessary for this response by the glands. The system is one of simple negative feedback.

Thyrocalcitonin

In spite of the multifarious actions of parathormone, it is indeed surprising that such a simple feedback system involving one hormone can control the calcium content of the blood in such a precise manner. Rasmussen pointed out that such a mechanism of control would be likely to lead to wide oscillations of plasma [Ca^{++}], whereas the concentration is remarkably stable. Sanderson showed that if the plasma [Ca^{++}] of normal dogs is artificially lowered by the injection of a calcium-chelating agent (ethylene-diamine-tetra-acetic acid), or is artificially raised by the injection of calcium salts, then the plasma [Ca]$^+$ returned to normal after a few hours. When these experiments were repeated on parathyroidectomized dogs, striking differences were seen. Following induced hypocalcaemia after parathyroidectomy the plasma [Ca]$^{++}$ failed to return to normal levels; this is readily explained by the lack of parathormone and therefore an inability to mobilize calcium from the bones. Of more significance was their finding that the infusion of calcium salts into parathyroidectomized dogs resulted in a prolonged hypercalcaemia which persisted for 24 hours. Now in a normal dog the speedy return of plasma [Ca^{++}] to normal levels used to be explained by a cessation of parathormone production. But the parathyroidectomized dog had no parathormone production, yet it still developed a prolonged hypercalcaemia after the injection of calcium salts. This type of observation has led workers to look for the presence of some *hypo*calcaemic factor arising from the parathyroid glands. Evidence for the presence of a new hormone capable of causing a lowering of plasma [Ca^{++}] was provided in 1961 by Copp and his colleagues. The hormonal factor was called calcitonin.

The experimental technique used was a type of cross circulation, and a simplified sketch of their technique is shown in fig. 7. Dogs were used in these experiments and their thyroid and parathyroid glands were removed from the body and placed in a warm, moist chamber and perfused with blood from the carotid artery of the experimental animal. After perfusing through the isolated glands the blood was returned to the animal by means of a canula inserted into the external jugular vein. Heparin was added to the blood to prevent clotting in the system. Side arms in the external circulation permitted the introduction of calcium ions, or alternately a calcium chelating agent, into the blood perfusing the isolated glands, and also permitted the removal of samples of blood which had just perfused through the isolated organs. This blood was used to determine its effect upon the calcium metabolism of another dog. It was found that, when the calcium content of blood perfusing the isolated glands was raised from the normal 10 to 12 mg %, then the blood leaving the isolated glands caused a fall in calcium content of the blood of the dog which had undergone the operation, or produced a similar effect when the blood was introduced into another dog. The hypocalcaemic

factor in the perfusate had a rapid onset of action (within 20 minutes)
and a brief duration of about one hour only.

More recent evidence indicates that the hypocalcaemic factor arises
not from the parathyroid glands but from the thyroid and it has been
renamed thyrocalcotinin. The hormone has been isolated from

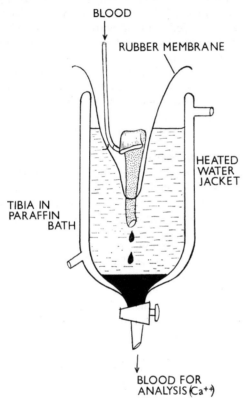

FIGURE 18. The effect of thyrocalcitonin on isolated perfused bone
(after Macintyre *et al.*, 1967).

thyroid tissue, purified and identified as a polypeptide. If the hormone is injected into an animal the hypocalcaemia which develops
cannot be prevented by removal of the kidneys or the gastrointestinal
tract. Thus the hypocalcaemia is not due to loss of calcium in the
urine or the gastrointestinal tract. It probably acts by causing a transfer of calcium from blood into bone, by increased calcification of bone
or by diminished decalcification, or by both mechanisms. A direct

action on bone has been demonstrated by perfusion studies of isolated bone. Fig. 18 shows a sketch of the bone perfusion technique used by Macintyre *et al.* (1967). A tibia was isolated from an animal, cleared of muscle, set up in a warmed bath of paraffin and perfused with blood by way of an artery. The upper end of the bone was occluded by a rubber sheath so that only blood leaving the lower end of the tibia was collected for analysis of calcium content. When thyrocalcitonin was added to blood perfusing the bone it produced a fall in the calcium content of blood leaving the bone, indicating an uptake of calcium by bone.

In summary we can say that the parathyroid and thyroid glands control the concentration of calcium ions in body fluids by the release of two hormones, having opposite effects, into the circulation and that the particular hormone released depends upon the concentration of calcium ions in the blood entering the glands.

Vitamin D and calcium metabolism

There are several substances which possess vitamin D activity. One such substance, D_3, can be manufactured in the skin from 7-dehydrotachysterol under the influence of ultra-violet light. D_3 synthesized in the skin is distributed throughout the body in the circulating blood, producing physiological effects on particular tissues, and as such fulfils the criterion of a hormone. However, endogenous supplies of D_3 are often insufficient because of coverings of the skin by clothes, because of inadequate sunlight, or because of atmospheric pollution which prevents the ultra-violet light reaching the skin in sufficient amounts. Thus there is an increasing dependence upon external sources of the substance and foods such as fish liver oils, eggs and butter are particularly rich sources. Therefore D_3 is also considered to be a vitamin.

The administration of vitamin D decreases the faecal excretion of calcium and phosphorus and this is due to an increased absorption of calcium from the intestine. The lowered phosphate excretion is the result of the decreased calcium excretion, which prevents the formation of insoluble calcium phosphate in the intestine.

This effect on intestinal absorption results in a rise in the concentration of calcium and phosphate in the blood. The parathyroid glands respond to the change in the composition of the blood by a reduced output of parathormone, and the urinary excretion of phosphate declines. Thus provided that there are adequate amounts of vitamin D, either from the skin or from the gut, sufficient calcium and phosphate are absorbed to provide for the growth and maintenance of bone. When supplies of vitamin D are deficient, supplies of calcium and phosphorus are no longer adequate for the calcification of bone and in the developing child the condition of rickets develops. However, the concentration of calcium in the blood is maintained at near-normal levels at the expense of the bones, and this is achieved by an increased production of parathormone. When vitamin D is administered in large doses, then it produces results which mimic those of parathormone itself—that is, there is a mobilization of calcium from bone and an increased renal excretion of phosphate. For this reason vitamin D can be used to treat hypoparathyroidism, a condition in which there is an inadequate production of parathormone. This condition can occasionally be produced by accidental removal of parathyroid tissue at thyroidectomy. Inadvertent treatment with vitamin D, for example by the over-addition of the vitamin to infant feeds, produces widespread disturbances of bone growth and calcium metabolism.

6

Hormones and Homeostasis— II. The Regulation of Glucose Metabolism

---

Introduction

In all mammals the central nervous system is almost entirely dependent on a supply of glucose for its metabolic needs. Many mammals, e.g. carnivores take in minimal amounts of carbohydrate in their diet, while others, e.g. man, ingest varying amounts of carbohydrate in their meals. Not all animals eat continuously throughout the day and some may go days without eating. For all these reasons there is a fluctuating supply of carbohydrate available from the intestine, often with long periods when no carbohydrate is available for absorption. A constant supply of glucose for the CNS is ensured by many complex mechanisms. These include emergency measures and also long-term measures. A rapid emergency measure is one in which glucose is liberated from glycogen stores. Long-term measures for maintaining glucose supplies involve the production of glucose from non-carbohydrate precursors, in particular from protein, a process which is called gluconeogenesis. These adaptive measures are brought about by the activation of particular enzyme systems or by the synthesis of increased amounts of certain enzymes. The regulation of the amount of glucose in the blood thus occurs at a molecular level and a variety of endocrine mechanisms co-operate in the process.

Factors involved in controlling blood glucose concentration

The amount of glucose in blood represents the balance between inflow of glucose into and the outflow of glucose from the blood. Cer-

tain organs such as the liver or the gut may add glucose to the blood circulating through them. The liver adds glucose to the blood in two ways:
(i) by the breakdown of glycogen,
(ii) by gluconeogenesis from protein.
The intestine can add glucose to the circulating blood by absorption of the products of carbohydrate digestion. Many organs, including the liver, can remove glucose from the blood to satisfy their own energy requirements, or for the formation of glycogen (e.g. liver, skeletal muscle, cardiac muscle, brain), or for the formation of fat. In view of the many possible pathways of glucose utilization and production, it is indeed remarkable that the concentration of glucose in blood remains relatively constant. In normal healthy young men the blood glucose rarely rises above 100 mg %, even after ingesting food rich in carbohydrate; only after a prolonged fast or very severe exercise does it fall below 60 mg %. This is only possible in a system where one member can add glucose to the blood if another removes glucose from the blood.

The rôle of the liver in glucose homeostasis

The liver plays a key rôle in glucose homeostasis because of its ability to discharge into the blood amounts of glucose which vary according to the needs of the organism. The liver is particularly important during fasting and at these times represents the major source of glucose to the body. The rôle of the liver is well demonstrated by the results of the experiments of Mann and his co-workers. They found that after removing the liver from a dog the level of glucose in the blood fell progressively and the animal eventually died. When a glucose solution was injected into a dog whose liver had been removed the animal lived until the glucose had all been used up.

There are two ways in which the liver contributes to blood glucose, by glycogenolysis and by gluconeogenesis.

(i) *Glycogenolysis*. This is an emergency system in which stores of glycogen can be used as a source of glucose (see p. 35). The supply of glycogen is of course limited, e.g. in man the total hepatic glycogen (about 75 g) is only sufficient to maintain the normal output of hepatic glucose for 12 hours. The enzyme phosphorylase is involved in the initial stage of glycogen breakdown and this enzyme is activated by a number of hormones (adrenaline, glucogon and ACTH) the secretion of which is known to increase whenever blood glucose concentration falls below a critical level (pp. 35, 109).

(ii) *Gluconeogenesis*. This describes the production of glucose from non-carbohydrate sources. The rôle of glucocorticoid hormones in hepatic gluconeogenesis is discussed fully in chapter 3. It must be appreciated that gluconeogenesis, unlike glycogenolysis, is not a rapidly acting emergency system and it takes several hours before glucocorticoid hormones can increase the rate of gluconeogenesis in the liver in response to a decrease in the amount of glucose in the blood.

Glucose metabolism in the liver is illustrated in fig. 19. Glucose-6-phosphate plays a central rôle in the metabolism, sitting as it does on all the routes of glucose utilization. The amount of glucose leaving the liver depends on a variety of factors, including the rate at which glucose enters the liver, the rate of glucose uptake by liver cells, the rate of conversion of glucose to glycogen and the rate of gluconeogenesis.

The rate of glucose uptake by cells of the liver

This depends on the rate of phosphorylation of glucose by the liver kinases. There is evidence that insulin can increase the phosphorylation of glucose and thereby increase the amount of glucose taken up by the liver cells, thus reducing the amount of glucose leaving the liver. Obviously this process of glucose phosphorylation must be be spatially separate from the enzyme glucose-6-phosphatase, otherwise as soon as glucose-6-phosphate was made it would be immediately dephosphorylated by glucose-6-phosphatase resulting in free glucose and heat production. Glucose-6-phosphatase in the intact cell is associated with the microsomes, separating it from the soluble phosphorylating systems.

The rate of gluconeogenesis

In this process glucose is derived from the catabolism of protein and other sources. The mechanisms which increase the rate of gluconeogenesis are triggered off during hypoglycaemia. Hypoglycaemia produces widespread activation of the sympathetic nervous system, which includes the outpouring of adrenaline from the adrenal medulla. This hormone, in addition to activating the hepatic enzyme phosphorylase and so producing a prompt increase in hepatic glucose output, also triggers off more slowly developing mechanisms which can increase hepatic glucose production. Adrenaline stimulates the anterior pituitary to secrete increased amounts of adrenocorticotrophic hormone which activates the adrenal cortex to synthesize and release increasing amounts of glucocorticoid hormones. These hormones by causing a release of lactate, various amino acids and three carbon precursors from various tissues, provide some of the raw materials for hepatic glucose production. In the liver the glucocorticoid hormones initiate an increase in the rate of synthesis of these intracellular enzymes associated with gluconeogenesis (p. 42). The adrenaline output which is an initial response to hypoglycaemia, also causes the release of glycerol from fat tissues, thus also providing more raw material for glucose synthesis.

In the absence of insulin, the liver will continue to secrete glucose into the blood, even in the presence of high concentrations of glucose in the blood (as high as 500 mg % in the dog), whereas in the presence of insulin the liver may show a net uptake of glucose from the blood even when blood glucose falls below 75 mg %. It seems, then, that the hormone insulin is an important factor in regulating glucose output from the liver. The hormone appears to be able to reduce the net glucose output from the liver by two mechanisms (fig. 19). First it can promote the uptake of glucose from the blood and its phosphorylation inside the liver cell, and this it may do by activating a specific glucokinase. Secondly, insulin may reduce net glucose output from the liver by depressing the rate of gluconeogenesis.

Thus the liver varies its output of glucose, depending on the requirements of the body. Information about body requirements is supplied in terms of various hormones. The output of hormones is in turn regulated by nervous processes, in particular those of the sympathetic nervous system, which are activated during hypoglycaemia.

Factors regulating the utilization of glucose by tissues

We have seen that the liver is a very important factor in the maintenance of adequate stores of circulating glucose for the metabolic needs of the central nervous system. Another important factor is the regulation of the utilization of glucose by various tissues.

In the presence of adequate dietary sources of glucose, many tissues are able to take up glucose from the blood for their own energy requirements, or for the synthesis of glycogen or fat. During periods when the dietary calorific intake is greater than the metabolic requirements, then large amounts of glucose may be diverted into triglyceride in adipose tissues. However, during periods of fasting these various routes of glucose utilization must be closed down to ensure that the limited supplies of glucose arising in the liver are available for the central nervous system. This is possible because muscle and fat tissue can utilize other substrates for their energy requirements, e.g. by oxidation of fatty acids and ketone bodies.

Various hormones are concerned in the modulation of glucose uptake by muscle and fat tissues in accordance with the state of nutrition. The rôle of insulin in regulating the cellular uptake of glucose is discussed fully in chapter 3. Many cells of the body, particularly muscle and fat cells, have as it were a cell membrane barrier to the

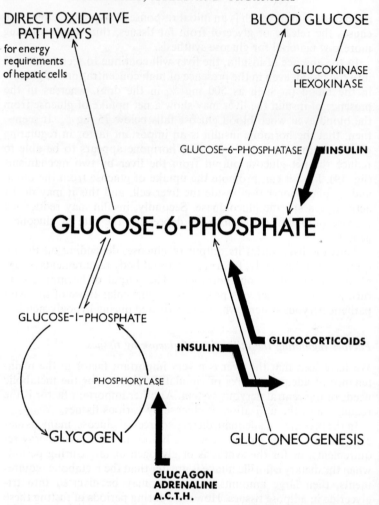

FIGURE 19. Glucose metabolism in the liver and the effects of hormones.

free movement of glucose into the cells from the extracellular fluid. Insulin acts by facilitating the transport of glucose across the cell membrane. Insulin production by the pancreas is at its greatest when the carbohydrate supplied in the diet is more than adequate to supply the requirements of the central nervous system. In this condition, excess glucose is diverted under the influence of insulin into muscle and fat tissues. Furthermore, when dietary supplies of carbohydrate are

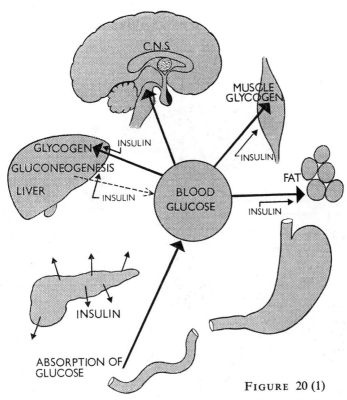

FIGURE 20 (1)

FIGURE 20. Endocrine control of glucose metabolism. 1. Illustrates metabolism of glucose during periods of dietary intake of excess carbohydrate. The secretion of insulin promotes the synthesis of hepatic and muscle glycogen, the synthesis of fats and depresses hepatic gluconeogenesis. 2. The change in the situation at the onset of hypoglycaemia. Insulin production declines and muscle and fat tissues are thus deprived of glucose. There is an outpouring of adrenaline from the adrenal medulla which acts at various sites to raise the blood sugar. (a) hepatic phosphorylase is activated; (b) there is a release of lactate from muscle and glycerol from fat tissues which act as raw material for hepatic gluconeogenesis; (c) activates, via the hypothalamus, the secretion of A.C.T.H. by the anterior pituitary gland. There is also secretion of glucagon from the pancreas which reinforces the action of adrenaline on hepatic phosphorylase. 3. Later stages in the response to hypoglycaemia. A.C.T.H. from the anterior pituitary gland activates the adrenal cortex which secretes glucocorticoid hormones. These hormones act in various ways to maintain supplies of glucose for the C.N.S.; (a) they antagonize the uptake of glucose by muscle and fat cells; (b) cause a release of amino acids from muscle which act as raw material for hepatic gluconeogenesis; (c) increase the amounts of those hepatic enzymes concerned in gluconeogenesis,

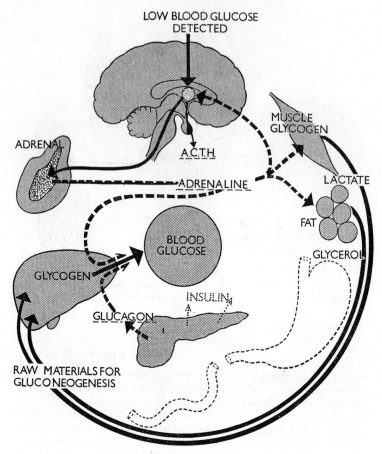

FIGURE 20 (2)

abundant, insulin can reduce the rate of hepatic glucose production by reducing the rate of hepatic gluconeogenesis (p. 75). When supplies of dietary carbohydrates are insufficient to maintain normal amounts of glucose in the blood, the output of insulin from the pancreas falls and muscle and fat cells no longer have free access to circulating glucose. The rate of hepatic gluconeogenesis, freed from the inhibiting effect of insulin and stimulated by glucocorticoid hormones, now increases.

Not all body tissues have a membrane barrier to the free movement of glucose into the cell. The cells of the central nervous system and the liver do not require insulin for the transport of glucose into the cell.

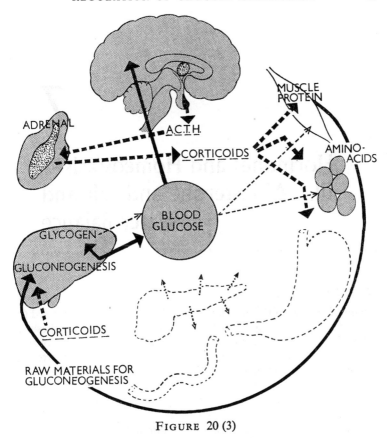

FIGURE 20 (3)

During fasting, insulin output is low and muscle and fat cells have to rely on sources of energy other than glucose from the blood. There is, moreover, a second defence mechanism which prevents the utilization of glucose by these tissues during fasting. This is achieved by the inhibitory effect of glucocorticoid hormones on the hexokinase re-action in these cells.

Summary

The regulation of the amount of glucose in the blood, to meet the needs of the central nervous system, is then a complex process, which is achieved by a system of activating and checking devices acting at various levels in the cells on the cell membrane, on enzyme systems (e.g. hexokinase) and on the genome (see fig. 20).

7

Hormones and Homeostasis— III. Aldosterone and Salt and Water Balance

----

The isolation of aldosterone

Chemical studies of the adrenal cortex of mammals have revealed the presence of approximately thirty different steroids. Many of these are probably intermediate compounds along the biosynthetic pathway to the active hormones released by the gland, or represent artefacts produced during the processes of chemical isolation. Of these compounds only seven have significant effects in maintaining life, in animals from which the adrenal glands have been removed. The amorphous non-crystalline fraction of extracts of the adrenal gland is also highly active in maintaining life in the adrenalectomized animal. In 1952 Tait, Simpson and their colleagues were led to apply the then new technique of chromatography to adrenocortical extracts and to blood from the adrenal vein in an attempt to identify the contents of the secretions of the adrenals. At this time there was difficulty in explaining the sodium-retaining action of the whole cortical extract in terms of the crystalline steroids that had been isolated from it. The new chromatographical methods led to the isolation of a new naturally occurring steroid called aldosterone, which had potent effects on sodium retention by the kidney.

The source of origin of aldosterone

Aldosterone is produced by the outermost layer of the cortex, the zona glomerulosa (plate 2). This is the only part of the cortex that remains structurally and functionally intact following the removal of the anterior pituitary gland. Pituitary adrenocorticotrophic hormone can influence the secretion of aldosterone from the glomerulosa, but the regulation of the activity of the glomerulosa appears to be more significantly under the control of other endocrine mechanisms (see below). By contrast the zona fasciculata and reticularis of the adrenal cortex undergo atrophy following the removal of the pituitary gland and a severe deficiency of glucocorticoid hormones develops.

Physiological actions of aldosterone

Following removal of the adrenal gland from mammals a variety of signs and symptoms arise which are attributable to a lack of the influence of both aldosterone and glucocorticoid hormones, such as corticosterone and hydrocortisone. The lack of aldosterone precipitates a critical loss of sodium from the body in the urine, and with it a loss of water. This leads to a progressive shrinkage of extracellular fluid volume and later of cellular fluid, unless free access to salt and water is available, or unless replacement therapy with mineralocorticoid hormone is given. Various adrenocortical and synthetic hormones exert some salt retaining action. Corticosterone and hydrocortisone promote some renal retention of sodium, especially when given in large doses, but before the discovery of aldosterone the substance desoxycorticosterone was the most potent compound in causing sodium retention. Desoxycorticosterone is a synthetic steroid and although it is present naturally in small quantities in the adrenal cortex, it is considered to be only an intermediary metabolite on the pathway to those hormones which are eventually released into the circulation, i.e. desoxycorticosterone is not a natural hormone. Aldosterone is known to be twenty-five to fifty times more active, on a weight basis, than desoxycorticosterone in causing sodium retention by the kidney.

Coincident with the effect of sodium retention by the kidney there is usually also an increased loss of potassium in the urine. It has been concluded that aldosterone stimulates an ion exchange mechanism in the renal tubules whereby sodium is absorbed from the glomerular filtrate in exchange for potassium, which is secreted into the urine. Hydrogen ions secreted into the urine by the tubule cells also compete for this mechanism of ion exchange and sodium may be reabsorbed in exchange for either potassium or hydrogen ions.

The site of action of aldosterone

Of the total amount of sodium which is filtered off in the glomeruli, perhaps only a few per cent, is reabsorbed under the influence of aldosterone. The bulk of the filtered sodium is actively reabsorbed in the proximal convoluted tubules of the nephrons. In the ascending limb of the loop of Henlé sodium ions are extruded from the lumen of the tubule into the peritubular fluid, thus creating a hypertonic intercellular fluid in the medulla. Reabsorption of water by the tubule depends on this hypertonicity of the intercellular fluid of the medulla. Only a small fraction of the filtered load of sodium eventually reaches the distal convoluted tubule of the kidney. Here, depending on the state of sodium balance and the levels of circulating aldosterone, further sodium can be absorbed in exchange for potassium ions which are secreted into the urine. This site of action of aldosterone was demonstrated using the technique of stop-flow analysis by Vander, Malvin, Wilde, Lapides, Sullivan and McMurray in 1958. In this technique the ureter of an animal is exposed for the collection of urine and an intravenous infusion is given containing, among other substances, the sugar mannitol. This sugar filters through the glomeruli but is not reabsorbed from the glomerular filtrate. Its presence in the tubules ensures osmotic retention of water and a high rate of urine flow therefore develops, i.e. mannitol causes an osmotic diuresis. When a high rate of urine flow has developed the ureter is clamped. This results in a rapid rise in intra-renal pressure, which by opposing the hydrostatic pressure in the glomerular capillaries, prevents new filtrate from being formed. Thus an essentially static column of fluid exists in the nephrons and each segment of the nephron can perform in an exaggerated fashion, those functions which it normally carries out to a lesser degree on a moving column of fluid. After an interval the clamp on the ureter is removed and the urine which gushes out under high pressure is collected in a large series of separate aliquots. It is reasonable to suppose that the first samples are from the collecting ducts, the second from the distal convoluted tubules and so on. Marker substances, injected intravenously before clamping the ducts also help in localizing the distribution of the aliquots. For instance, para-aminohippurate which is secreted into the urine from the proximal tubule, helps to assign those aliquots coming from this part of the nephron. By determining the sodium content of these separate aliquots, one can discover the pattern of sodium handling along the length of the nephron. By comparison of the findings in the presence and absence of injected aldosterone, one can deduce the effect and site of action of the hormone.

In addition to this renal effect of aldosterone there are effects on other tissues. Thus administration of aldosterone results in a fall in the sodium content, and an increase in the potassium content of saliva, sweat and faeces, and attempts have been made to use changes in the sodium:potassium ratio of saliva as an index of the amount circulating aldosterone.

The regulation of aldosterone production

A variety of experimental disturbances of electrolyte metabolism and cardiovascular function have been shown to be associated with changes in aldosterone production by the adrenal cortex.

The effect of changes in the concentration of sodium and potassium ions in the plasma

In studying the mode of action of changes in the electrolyte balance on adrenocorticoid function, McDonald, Goding and Wright (1958) used an ingenious technique of auto-transplantation of the adrenal gland. One adrenal gland was translated to the neck with vascular anastomoses to the carotid artery and the jugular vein. The vessels and the transplant were included in a flap of skin which permitted local infusion of substances into the blood supply of the adrenal to be made. These infusions altered the electrolyte content of the blood perfusing the adrenal gland. The technique also permitted easy sampling of adrenal venous blood for assay of its aldosterone content. The other adrenal gland was then removed so that the animal was entirely dependent on the transplanted adrenal. Sheep were the animals used in these experiments. In ruminants, large amounts of sodium are secreted in the saliva and if the duct of the parotid salivary gland is drained to the exterior via a fistula, thereby preventing the saliva from being swallowed, then a state of negative sodium balance develops. As much as 500 milli-equivalent of sodium per day can be lost in saliva. The sodium state of the animal can then be adjusted by giving the sheep saline solutions to drink. It was found that in a state of normal sodium balance, basal levels of aldosterone as measured from the transplanted gland were in the order of 0–0·48 μg per hour. When sodium loss was allowed to exceed intake, the kidney responded by an increased reabsorption of sodium. At this stage, aldosterone production was between 1 and 3 μg per hour. With increasing losses of sodium the aldosterone output increased further to 15 μg per hour. At this stage salivary secretions showed a reduced sodium and an increased potassium content. Thus the degree of sodium deficiency determines the amount of aldosterone secreted.

HOW CHANGES IN PLASMA ELECTROLYTES MODIFY ALDOSTERONE PRODUCTION

Investigations were carried out in order to try to determine the mechanism by which sodium deprivation stimulates aldosterone production. One obvious possibility was that the adrenal gland is itself sensitive to changes in plasma sodium concentration. This possibility was tested by giving local infusions of low sodium concentration into the isolated adrenal gland. This procedure proved ineffective as a stimulus for aldosterone secretion, unless the fall in the concentration of sodium was also associated with a rise in the concentration of potassium. Infusion of sodium-rich solutions into the isolated adrenal gland of an animal which was already responding to sodium deprivation by increased aldosterone production, did *not* cause a reduction in the rate of release of the hormone. It appears, then, that sodium deprivation is sensed in some other part of the body, which in turn activates the adrenal cortex. That this activation is by means of a humoral agent, was illustrated by cross-circulation experiments. An adrenalectomized sheep was subjected to sodium depletion and then the blood from this animal was cross-circulated to the adrenal gland of a normal sheep which was in sodium balance. This cross circulation resulted in an increased secretion of aldosterone in the recipient animal. This type of experiment demonstrates the presence, in the circulatory blood of sodium-depleted animals, of some circulating humoral agent which can stimulate the production of aldosterone by the adrenal cortex. We can designate this substance as ASH (aldosterone-stimulating hormone).

A variety of tissues have been removed from animals in the hope of finding the source of origin of ASH. Removal of the pituitary, the pineal, the subcommisural body, do not prevent the adrenal response to sodium deprivation nor does the removal of the nervous system anterior to the mid-collicular level.

The effect of changes in cardiovascular function on aldosterone production

Various types of interference with cardiovascular function have been associated with an increase in aldosterone production. Thus haemorrhage, constriction of the inferior vena cava producing a fall in central venous pressure, and reduction of the carotid artery pulse pressure have all been reported as causing an increase in aldosterone production. The presence of a humoral agent stimulating aldosterone production was shown by cross-circulation experiments in dogs with thoracic caval obstruction. The blood of the dogs with the caval obstruction, when circulated through normal dogs (with isolated adrenal glands) caused an increased secretion of aldosterone in the normal dogs.

Identification of possible sources of ASH were then tested by the classical technique in which various tissues were removed prior to haemorrhage and estimation of aldosterone production. Removal of the anterior pituitary gland caused a fall in aldosterone production because of the loss of ACTH. It is known that ACTH does stimulate aldosterone production and that ACTH release from the pituitary increases in response to haemorrhage. However, a substantial increase in aldosterone production in response to haemorrhage still occurred after hypophysectomy. Thus there must be sources of ASH other than the pituitary. In addition to hypophysectomy, removal of either the liver, or decapitation, reduced the aldosterone response to haemorrhage. Removal of the kidneys caused a 50% reduction in aldosterone production after haemorrhage. However, after hypophysectomy, nephrectomy *completely* prevented the adrenal response to haemorrhage. This implicates the kidney as a potent source of ASH and this is confirmed by observations on the effects of injections of saline extracts of kidney, which produced a sevenfold increase in aldosterone production.

RENIN, ANGIOTENSIN AND ALDOSTERONE PRODUCTION
A close relationship between the kidney and the adrenal cortex has been suggested by many workers. A substance called renin was isolated from the kidney in the late 19th century by Tigerstedt and Bergman. It was shown that renin is a proteolytic enzyme and that it reacts with a plasma substrate to produce a substance called angiotensin. The plasma substrate which reacts with renin is an alpha-2 globulin and the product of the proteolytic activity of renin is a weakly vasopressor substance, a decapeptide called angiotensin I. This decapeptide is then converted into a highly active pressor octapeptide called angiotensin II.

In 1951 Deane and Masson reported that in rats the injection of a purified solution of renin produced an enlargement of the zona glomerulosa of the adrenal cortex. Further work has produced very strong evidence that renin is produced by the juxta-glomerular apparatus in the kidney. For example, microdissection studies of kidneys have shown that most or all of the renin is contained in the juxta-glomerular area. In 1961, using the fluorescent antibody technique (p. 15), Edelman and Hartcroft localized renin to the juxta-glomerular cells.

THE STRUCTURE OF THE JUXTA-GLOMERULAR APPARATUS
The juxta-glomerular cells are located in the wall of the afferent arteriole of the glomerulus. The smooth muscle cells of the media of the afferent arteriole gives way to large round, or polygonal, epithelioid cells as the afferent arteriole approaches the glomerulus. This tissue forms a cuff in the afferent arteriole which butts against the wall of Bowman's capsule. The cells contain many granules which stain brilliantly with fuchsin. This perivascular cuff of cells is called the 'polkissen' or polar cushion. In association with this polar cuff there is a special area of the distal-convoluted tubule closely applied to the afferent arteriole. In a localized area of the distal-convoluted tubule the cells are high columnar, rather than cuboidal, and they

contain large nuclei. This area is called the macula densa and together with the polar cuff it forms the juxta-glomerular apparatus. There is at this moment no evidence which implicates the macula densa in electrolyte metabolism, but it is described here for the sake of completeness.

Various workers have provided evidence which suggests that the ASH secreted by the kidney is in fact renin and that the renin-angiotensin system is important in regulating the secretion of aldosterone by the adrenal cortex. Intravenous infusion of either renin or synthetic angiotensin II, stimulates aldosterone secretion. Angiotensin II acts directly on the adrenal cortex and marked increases in aldosterone secretion followed the injection of small amounts of angiotensin (less than 0·5 μg.) into the artery of the isolated adrenal gland.

Fractionation studies of crude kidney extracts showed that the only fraction having aldosterone stimulating properties was renin. In hypophysectomized dogs, which were producing increased amounts of aldosterone in reponse to chronic sodium depletion, thoracic caval obstruction or congestive heart failure, the removal of the kidneys was followed by a marked reduction in aldosterone production.

Summary of the renal control of aldosterone production

The following is a scheme suggested by Davies for the control of aldosterone secretion. He lists the following components of the renal aldosterone regulating system.

1. The renal afferent arterioles. These form the 'volume receptor' and the signal to these may be a decrease in the amount to which the renal afferent arteriole is stretched.
2. The juxta-glomerular releases renin when there is a decrease in stretch of the renal afferent arterioles.
3. The liberation of renin from the kidney results in the formation of angiotensin II in the plasma.
4. Angiotensin II stimulates the zona glomerulosa of the adrenal cortex to produce more aldosterone.
5. Aldosterone is transported in the blood stream to the kidney where it promotes retention of sodium.
6. The renal retention of sodium, and indirectly of water, causes expansion of the blood volume, a rise in blood pressure and an increased renal blood flow. The original stimulus to the kidney, which was a fall in pressure in the renal afferent arterioles, is now removed and the release of renin declines, i.e. a negative feedback mechanism exists.

This scheme is illustrated diagrammatically in fig. 21.

animal which is in normal fluid and electrolyte balance. When aldosterone is administered to an animal there may be considerable delay before there are demonstrable effects on the electrolyte balance and excretion. In man this delay is about 2 hours, but it may be as long as 6 hours. Furthermore, if one administers aldosterone to an animal over a long period of time the effect of the hormone diminishes. After the initial sodium and water retention and the loss of potassium, there is a gradual 'escape' from these effects and sodium excretion slowly increases again. The mechanism of this 'escape' is unknown. It has been doubted by some workers whether the small quantities of aldosterone which circulate can be sufficient to play a significant rôle in the regulation of sodium and potassium excretion. The quantity of aldosterone produced by a normal man in normal fluid and electrolyte balance amounts to 150–300 µg/day. The concentration of aldosterone in blood plasma has been estimated to be about 0·5 µg/100 ml.

The delay in the onset of the renal effects of aldosterone certainly means that the hormone is unlikely to be concerned in the minute to minute regulation of sodium excretion. Rapid changes in sodium excretion do occur when the position of the body is altered, e.g. from lying down to standing up, when small amounts of blood are lost, when venous congestion is produced, or when large volumes of isotonic fluids are infused. These changes in salt excretion may be accompanied by changes in renal blood flow or in the rate of glomerular filtration, but they may also occur in the absence of any measurable changes in these parameters of kidney function. Neither do these changes require the presence of a functional adrenal cortex.

It has been suggested that the hypothalamus may be a possible target organ for the afferent 'volume' signal patterns which arise during the above interferences with cardiovascular function. It has been known for many years that lesions in the caudal brain stem may be followed by a loss of electrolytes in the urine. More rostral brain damage is the basis of a disease called 'cerebral salt wasting' a condition in which there is a serious loss of electrolytes in the urine, which is of such a degree that there may be a gross depletion of extracellular fluid. The administration of aldosterone will not always bring these patients into electrolyte balance—only salt substitution is effective. In the rat, experimental lesions in the region of the posterior hypothalamus produces a similar condition. Available evidence suggests that the posterior hypothalamus exerts a tonic influence in maintaining the level of sodium reabsorption by the kidney. This 'antinatriuretic' centre can be inhibited by some pattern of baroreceptor signals from the thoracic vascular bed. From experiments involving the infusion of 0·9% saline intravenously into dogs, de Warner and his colleagues (1961) were led to conclude that the sodium diuresis which followed was due to the presence of a circulatory hormone, which was neither vasopressin nor aldosterone. The source of the hormone is unknown, but the baroreceptors involved in the reflex appeared to be in the pulmonary circulation.

It is against this background that we must assign the position of aldosterone in electrolyte and fluid homeostasis. The body appears to possess a variety of mechanisms which regulate the volume and composition of extracellular fluid. The position of aldosterone would appear to be that of a reserve mechanism, which is called into action when the integrity of homeostasis is threatened by reduction of fluid or electrolyte intake, or by haemorrhage or disease.

D

8

Hormones and Homeostasis— IV. The Control of Water Balance

The relationship between salt and water balance

Body water balance is achieved by a variety of mechanisms, both nervous and humoral, which regulate water intake and water losses. Water intake is regulated by nervous mechanisms acting at hypothalamic and higher levels of organization, and directed via the sensation of thirst, at maintaining a normal osmolarity of body fluids. Water losses occur in various regions of the organism. Losses of water by way of the expired air and evaporation through the skin represent inevitable losses in that they are not under the control of the body's water-regulating mechanisms. However, this is not true of the water lost in the urine. A certain minimal water loss is necessary in order to excrete nitrogenous and other waste products of metabolism. However, when these requirements have been met, the kidney can vary the volume of water in the urine and can excrete small amounts of concentrated urine or copious amounts of dilute urine. It is the humoral regulation of this facultative fraction of water losses that we shall be discussing in this chapter.

The situation is complicated by the fact that the regulation of the sodium balance also indirectly determines the water balance. Sodium is the major osmotically active substance in the extracellular fluid, and a gain in sodium is associated with a gain in water; a loss of sodium with a loss of water. These changes are associated with the necessity of maintaining a normal osmolarity of body fluids. There are separate regulating mechanisms for the control of sodium losses and water losses, so that to some extent these substances can be handled separately and one or the other may be preferentially excreted or retained. However, it often happens that the mechanisms for the control of sodium and of water are activated simultaneously. Thus after the loss of a significant amount of blood, separate endocrine mechanisms for both salt and water retention are activated simultaneously in order to conserve extracellular fluid of a normal composition.

Antidiuretic hormone ADH or vasopressin

There is clear evidence that the posterior lobe of the pituitary gland, the pars nervosa, is closely concerned with the homeostatic regulatory mechanisms for water. Pathological or experimental destruction of the pars nervosa results in a condition known as diabetes insipidus, in which large volumes of dilute urine are produced by the kidneys, necessitating the drinking of copious amounts of water in order to maintain water balance. The condition can be relieved by the injection of extracts of the pars nervosa, or even by the application of dried extracts to the nasal mucosa, the active principle being absorbed into the blood stream through the nasal mucous membrane.

Secretion of ADH

The pars nervosa is richly innervated with non-myelinated nerve fibres which have their origin from cell bodies situated mainly in the supra-optic and para-ventricular nuclei of the hypothalamus (fig. 5). The gland also contains various types of cells, some of which are called pituicytes whose cytoplasm is rich in granules and vacuoles. Until fairly recently it was thought that these pituicytes (undifferentiated neurological cells) were the source of the humoral factor that can be extracted from the gland. Recent experimental work indicates that the hormones stored in the pars nervosa (oxytocin and ADH) are actually produced by modified nerve cells called neurosecretory cells. These cells lie in the pre-optic and paraventricular ganglia of the hypothalamus. The neurosecretory material migrates down the axons of these neurones and is deposited in the dilated axon terminals which are closely applied to blood capillaries in the pars nervosa. The neurosecretory material is transported down the axons in association with a 'carrier' protein in the form of granules. Distal migration of these granules has been observed in the living pituitary stalk. If the bundles of axons connecting the hypothalamus and the pars nervosa (i.e. the hypothalamico-hypophyseal tract) is sectioned, then it is possible to demonstrate a progressive accumulation of neurosecretory material proximal to the section. Further confirmation of this migration of neurosecretory granules was obtained using the technique of autoradiography, by injecting a radioactive precursor of the hormones into the cerebral ventricles (p. 16).

The active principles which can be extracted from the posterior lobe are peptides having 8 amino acids in the molecule and a molecular weight of about 1,000. These peptides are extracted in association with a protein of molecular weight 30,000. ADH and oxytocin can be removed from combination with the protein by procedures which do not rupture peptide bonds, so that it is unlikely that the peptides are fragments of the protein. It seems likely that about sixteen of these peptide molecules are loosely bound to each carrier protein molecule.

The hormones can be released rapidly from the nerve terminals in response to appropriate stimuli. Release is regulated by nerve impulses passing down the nerve fibres of the hypothalamico-hypophyseal tract from the hypothalamic nuclei. There is no choline acetylase (the enzyme concerned in the synthesis of acetylcholine) in the pars nervosa, and it is unlikely that acetylcholine is a mediator in the release of the hormones. Appropriate stimuli usually cause the release of both oxytocin and vasopressin into the circulation. Only with recent methods of assay of hormones (1967), has it been possible to establish that one hormone can be released in the complete absence of the other. In man at least, oxytocin seems to have no effect on renal function. In the human male the significance of oxytocin is unknown, although in the female it is known to be involved in the 'let down' of milk and in uterine activity in labour. Antidiuretic hormone produces a retention of water by the kidney and thereby prevents the production of a large volume of urine (i.e. it opposes diuresis). The alternative name for ADH is vasopressin, and this name describes an action of the hormone on blood pressure. The effect on blood pressure is observed only after the administration of large non-physiological doses of the hormone. The name of antidiuretic hormone is preferable because it describes its primary physiological action.

Mechanisms regulating ADH secretion

Stimuli which induce changes in the rate of secretion of ADH from the pars nervosa act by way of the hypothalamus. This is indicated by the effects of pituitary stalk section, or destruction of the hypothalamic nuclei. These procedures give effects similar to those of removal of the pars nervosa itself, with the exception that the ensuing diabetes insipidus (gross losses of water in urine and associated thirst) is of a more severe type than that following hypophysectomy. Electrical stimulation of the appropriate hypothalamic nuclei results in an increased rate of release of ADH from the pars nervosa.

OSMOTIC STIMULI

Verney showed that the injection of hypertonic solutions of salt or sucrose into the carotoid artery resulted in an increased liberation of ADH, which could be detected in jugular vein blood. As small a change as a 2% increase in osmotic pressure of blood produces a prompt release of ADH. A fall in the osmotic pressure of blood, produced for example by the intake of water, suppresses release of ADH from the pars nervosa and the kidney responds to the declining amounts of ADH in blood by producing a more copious flow of urine. There is a time lag before these effects are noticed and this is accounted for by the need to inactivate or excrete the ADH already circulating in the blood. For this reason it is difficult to demonstrate the effect of small falls in osmotic pressure of blood on ADH secretion and flow of urine.

BARORECEPTORS

The rôle of the kidney in the regulation of the volume of body fluid, was indicated by Peters (1935). He stated that, 'the fullness of the blood stream may provoke the diuretic response on the part of the kidney'. Following a loss of body fluids, e.g. in haemorrhage, renal compensation produces a retention of both sodium and water in an 'attempt', to ensure the conservation of the volume and composition of body fluid. Two endocrine mechanisms act independently, but synergistically, in this conservation, first the release of ADH promotes water retention, and secondly, aldosterone promotes sodium retention (chapter 10). In this chapter we are concerned with the first of these two mechanisms. Bleeding to the extent of 9 ml/kg in man produced a retention of both salt and water by the kidney. If only 7 ml/kg was taken (with the patient recumbent) only the amount of water excreted was altered. This response is explained by an increased rate of release of ADH from the pars nervosa. Share in 1962 demonstrated a sixfold increase in the ADH content of the jugular vein blood 20 minutes after a reduction of plasma volume by 8%. From this evidence one can say that the first system to be activated after a moderate change in blood volume is the ADH system. There is an increased rate of secretion of ADH and the initiation of anti-diuresis. Only after more severe changes in blood volume is the mechanism which activates sodium retention by the kidney, brought into action.

This type of reflex response to changes in volume of body fluid requires the presence of pressure-sensing elements (baroreceptors) in the vascular system. To anticipate somewhat, once can say that baroreceptors are present in various components of the vascular tree.

Baroreceptors in the low-pressure compartment (veins) are probably those concerned in the regulation of ADH secretion, whereas baroreceptors in the afferent glomerular vessels, in the high-pressure compartment (arteries) are concerned with the regulation of aldosterone secretion. There are also the classical baroreceptors in the carotid artery and aortic arch and these are intimately concerned with initiating cardiovascular responses to changes in blood volume and pressure. It has been suggested that there may also be receptors in the interstitial spaces of tissues since changes in blood volume may produce secondary changes in the volume of interstitial fluid.

Attempts to locate the distribution of baroreceptors concerned with fluid volume control have involved a great variety of experimental techniques. These include a study of the effects of haemorrhage, transfusion, pressure breathing, the immersion of the whole body in water, changes in posture, occlusion of venous drainage from the limbs, stretching of the chambers of the heart and so on. These procedures have been accompanied by simultaneous studies of renal function to detect changes in the handling of sodium, and assays of blood, to detect any humoral factors such as ADH or aldosterone. The situation is complicated by the fact that not all the changes in renal function which follow changes in blood volume can be accounted for by either ADH or aldosterone. However, we will study some of the relevant evidence which implicates ADH in the regulation of blood volume.

Experiments designed to demonstrate the effects of haemorrhage on ADH release and on urine flow have already been described. Further evidence for the rôle of ADH in the regulation of blood volume comes from studies of the effects of negative and positive pressure breathing. X-ray studies have provided the following data on blood distribution:

Negative pressure breathing	*Positive pressure breathing*
Blood is displaced into the thoracic veins.	There is depletion of blood in the thoracic veins.
Increases in the loss of water from the kidney.	A decreased amount of water is lost from the kidney (oliguria).

The changes in the amount of urine produced in these experiments have been shown to depend upon the amount of ADH secreted by the pars nervosa. It is significant that positive breathing does not produce oliguria when alcohol is present in the blood, for alcohol is known to effectively block the secretion of ADH from the pars nervosa. These results suggest that alterations in the rate of ADH releases are a result of changes in the stimuli received by baroreceptors located in the thoracic bed.

Experiments designed to localize these baroreceptors have involved the production of congestion in particular parts of the thoracic venous system. This has been done by the manipulation of loops tied around the pulmonary veins, so producing a rise of pulmonary venous pressure, or by inflation of a balloon in the left atrium to obstruct the flow of blood through the mitral valve, so increasing the pressure in the left atrium and the pulmonary veins. Experiments such as these have produced data which suggests that there is a sensitive baroreceptor region between the extrapericardial part of the pulmonary veins and the mitral valve. Other work suggests that similar baroreceptors, which can influence ADH release, exist in the right ventricle also. These receptors have been found to transmit information to the central nervous system via the vagus nerve. Cooling the vagus nerve to 8° C, blocks the conduction of impulses and also prevents the renal response to negative pressure breathing or to the inflation of a balloon in the atrium. The course taken by the nerve impulses from the atrial baroreceptors has been worked out. The vagus ganglion is connected to the central nervous system by a series of rootlets and electrical recordings of the most anterior (rostral) of these rootlets demonstrated discharges of nerve impulses which are synchronous with the heart-beat. When a nerve fibre is cut degeneration of the fibre proceeds to the cell body of the fibre and when these anterior rootlets are cut a degeneration of nerve-cell bodies occurs in a small area near the sensory nuclei of the vagus nerve. The path taken by the impulses from this point is yet to be worked out, but it is presumed that they are relayed by multisynaptic pathways in the periaqueductal grey matter to the mid-brain limbic area, where there are connections to the supra-optic area of the hypothalamus which controls the release of ADH.

Carotid sinus baroreceptors

The classical baroreceptors in the carotid sinus are also involved in the regulation of the secretion of vasopressin (Share and Levy, 1962, Clarke and Rocha e Silva, Jr, 1967). In the dog, reduction of blood pressure in the carotid sinus produced by compression of the carotid arteries results in a release of vasopressin. This effect is abolished if the sinus nerves are divided. Similar findings have been obtained in the cat. If both vagi and sinus nerves are divided in these animals, then there is a substantial reduction in the release of vasopressin in response to haemorrhage. Conversely, if the vagus nerve is electrically stimulated, then there is a release of vasopressin. However, vasopressin release occurs only if the blood pressure falls at least by 80 mmHg. This type of experimental finding has led to the view that the release of vasopressin during haemorrhage may be related more to a fall in blood pressure than to changes in blood volume.

Carotid body chemoreceptors

The chemoreceptors of the carotid bodies are also concerned in the response of the animal to haemorrhage. These structures contain elements sensitive to changes in pp CO_2, pp O_2 and [H^+] in the circulating blood. Carotid artery occlusion or haemorrhage results in

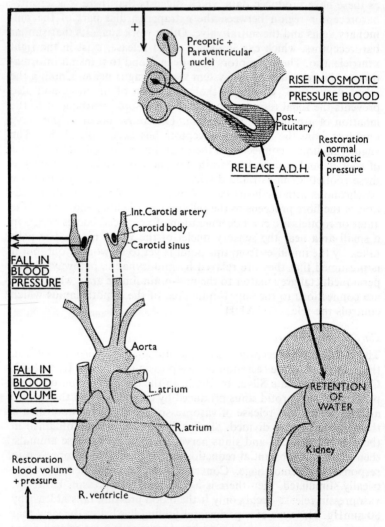

FIGURE 22. Summary of the control of release of A.D.H. from the posterior pituitary gland.

anoxia of the carotid bodies and increased rate of firing of the sensory endings. This stimulation of the carotid bodies initiates important responses of the cardiovascular system and ventilation. In addition there is a stimulation of the release of vasopressin. Share and Levy (1966) stimulated carotid body chemoreceptors by perfusing the carotid sinus with deoxygenated blood and noted an increased release of vasopressin.

The various afferent nerves from the atria, carotid sinus and carotid bodies appear to contain both inhibitory and excitatory influences on the release of vasopressin. Thus the hypothalamus may be under a tonic inhibitory influence from aortic and atrial baroreceptors, suppressing the release of vasopressin. This would explain the increased release of vasopressin associated with reduced baroreceptor discharge after haemorrhage, carotid occlusion and section of the vagi and sinus nerves. It would also explain the fall in vasopressin release after increased baroreceptor discharge caused by distention of the atrium. The activation of chemoreceptors by anoxia (due to carotid occlusion or haemhorrhage with hypotension) would provide, in addition, a positive influence on the release of vasopressin.

The actions of anti-diuretic hormone

The kidney is able to vary its ouput of water and solute according to the state of the salt and water balance. When an animal is deprived of salt, the urine may contain only a few milli-equivalents per litre of salt (cf. plasma [Na] = 120 m.equiv./l). When salt intake is high the concentration of salt in the urine can increase to twice the plasma values. When water intake is high, the urine is copious and dilute with osmotic pressures as low as one-tenth of the osmotic pressure of the plasma (plasma o.p. = 300 m.osmols/l). When an animal is deprived of water the osmotic pressure of the urine may rise above 1,200 m.osmols/l. This variation in osmotic pressure of urine is regulated principally by changing the amounts of ADH released from the pars nervosa.

It is not possible to present here a comprehensive account of the mechanism of urine formation and we shall only indicate those renal mechanisms which are influenced by ADH to provoke the absorption of water from the glomerular filtrate so as to produce a concentrated urine. The basis of the kidney's concentrating mechanism is the presence in the interstitial spaces of the medulla of a fluid hypertonic to blood plasma. In fact, there is a gradient of hypertonicity from the cortico-medullary junction to the tips of the pyramids. At the cortico-medullary junction the interstitial fluid bathing the nephrons and blood vessels is isotonic to blood plasma (300 m.osmols/l), and increases progressively to the tips of the pyramids where the osmotic pressure of the interstitial fluid may be 1,200 m.osmols/l, or more depending on the species of mammal. This gradient is generated by the active extrusion of sodium ions, unaccompanied by water, from the filtrate in the ascending limb of the loop of Henlé, into the interstitial fluid. This effect is multiplied by the countercurrent flow in the loop of Henlé. For details of this mechanism a textbook of general physiology should be consulted. The hypertonic medullary interstitial fluid is conserved by a low absolute rate of blood

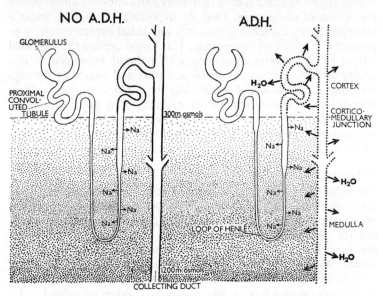

FIGURE 23. The effect of A.D.H. on the water permeability of the nephron. The driving force for water reabsorption is shown as an osmotic gradient in the medulla.

flow in the renal medulla. This hypertonic medullary interstitial fluid provides the driving force for water reabsorption. Provided that the distal convoluted tubules and collecting ducts are permeable to water, then the fluid in these tubules can osmotically equilibrate with the interstitial fluid. Water diffuses into the interstitium and concentration of urine results. The water permeability of these parts of the nephron is regulated by ADH. In the absence of ADH they are virtually impermeable to water and the urine produced is copious and dilute. In the presence of increasing amounts of ADH the distal-convoluted tubules and collecting ducts become increasingly permeable to water and concentration of the urine results (fig. 23).

The mode of action of ADH

It is difficult to study the mode of action of ADH on an organ as complex as the kidney. However, by means of micropuncture of nephrons, catheterization of collecting ducts and the technique of stop-flow analysis (p. 82) it has been possible to demonstrate one site of action of the hormone on water absorption. In addition to an action of ADH in promoting an increase in the permeability to water of the distal-convoluted tubules and collecting ducts, there may be further sites of action. It has been suggested that ADH may also reduce medullary blood flow and increase the rate at which sodium is extruded from the ascending loop of Henlé. These actions would help to promote and conserve the hypertonic medullary interstitium on which water reabsorption depends.

ADH has effects on other organs in other species. It increases permeability of frog skin and toad bladder to water. These structures

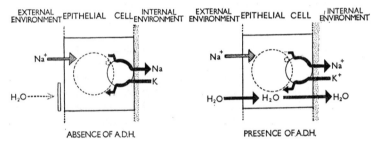

FIGURE 24. The effect of A.D.H. on amphibian skin.

are much more amenable to experimental investigation of the mode of action of ADH, than is the mammalian kidney. Stretches of skin can be suspended between fluid-filled chambers and the movement of substances from one chamber to another can be measured. Bladder can be readily converted into a closed sac which can be readily suspended in appropriate fluid media; water uptake can be readily measured by noting changes in weight of the sac.

Unlike many body cells, the cells of the amphibian skin or bladder are polar, that is they are not surrounded by a cell membrane which is uniform in its properties throughout its surface, but by a cell membrane which presents two distinct barriers to the diffusion of ions. The external barrier separates the cell contents from the environment, either the surrounding water or marshland in the case of the skin, or from urine in the case of the bladder. The internal barrier separates the cell contents from the interstitial fluid of the tissue. Fig. 24 shows some of the characteristics of these membranes. The inner membrane is regarded as being permeable to potassium but much less permeable to sodium. Sodium is actively extruded from the interior of the cell into the interstitial fluid. The 'pump' which extruded these ions is probably a potassium-linked one, for every sodium ion discharged from the cell one potassium ion enters the cell. By this mechanism the cell can maintain the high internal concentration of potassium which is characteristic of all cells. The internal barrier is permeable to water, since it has been shown that the cell volume changes in response to shifts of osmotic pressure of the interstitial fluid bathing the cell.

The outer membrane is impermeable to water despite the fact that it is permeable to sodium (and not to potassium). These characteristics permit the uptake of sodium from pond water or urine during hibernation, when the frog ingests no food but when sodium is being continuously excreted by the kidney. The impermeability of the outer membrane to water prevents the continued uptake of water, which would swamp renal function. The animal would behave as an osmometer and the tissues would be flooded by water entering via the skin, if the membrane were permeable to water.

The addition of ADH to frog skin, or bladder, produces striking effects on its permeability characteristics. As early as 1921 it was observed that there was a rapid swelling after the administration of the neurohypophyseal hormone to the intact frog. This observation has been repeatedly confirmed. There is not only an increased movement of water across the skin, but also an increased transport of sodium across the skin, from the external surface to the internal tissue fluid. The increase in the amount of sodium transported across the membrane, first noted in 1951 by Fuhrman and Ussing, could be due to a stimulation of the sodium pump, or to an increase in the sodium permeability of the outer membrane, thus permitting more sodium to reach the pump. This last hypothesis can explain the action of ADH on water and sodium movements on the basis of a single action of the hormone on the outer membrane. However, several workers believe that the effect on sodium and on water are independent effects of the hormone.

Action of ADH on cyclic adenosine monophosphate

It has been suggested that cyclic adenosine monophosphate (AMP) plays a key rôle in the effect of several hormones acting on different tissues. Glucagon (p. 36), ACTH and adrenaline (pp. 122, 114) all stimulate the production of cyclic AMP in their target organs; indeed, the addition of AMP to these tissues can mimic the effect of the hormones. In 1959 Hilton and his colleagues showed that ADH could mimic the effect of ACTH on the adrenal gland, and that of glucagon on the liver. In view of this it was suggested by Orloff and Handler, that ADH might stimulate the formation of AMP in its target tissues (amphibian skin and bladder), and the AMP could then initiate the changes in transport of sodium and water. Some evidence for this idea was provided. Support came from studies on the metabolic turnover of cyclic AMP. In the tissues this is degraded to 5-AMP, but this step can be inhibited by various substances including theophylline. It was found that either inhibition of degradation of 3',5'-AMP, the application of vasopressin (ADH), or the application of 3·5-AMP itself produced very similar effects on water movement and sodium transport by the toad bladder.

Support for a common mode of action of ADH in amphibian epithelia and mammalian kidney stems from observations by Brown and his colleagues (1963), who demonstrated a similar stimulation of cyclic AMP production by ADH in the kidney of a dog. It would seem highly probable that ADH produces its effects in these tissues by similar mechanisms. However, Ginetzinsky (1958) proposed that in the mammalian kidney ADH produces its effect by causing the secretion of an enzyme hyaluronidase by the kidney tubule. This enzyme, it was suggested, acts by hydrolysis of the hyaluronic acid on the surface of tubular cells, and the intercellular cement substance, thereby causing an increased permeability to water. This hypothesis was derived from observations that during antidiuresis there was an increased concentration of hyaluronidase in the urine. However, when hyaluronidase was applied to the isolated toad bladder, even in large amounts, it produced no changes in water permeability or in sodium transport.

It is obvious that the mode of action of ADH on these various structures is still a matter of debate. Even if one accepts the more likely explanation of an action through activation of cyclic AMP production, this is still several steps away from the observed changes in membrane permeability. It has been suggested that ADH increases the radius of certain pores in the membrane of the target cells, but the link between this proposed change and the production of cyclic AMP awaits elucidation.

9

Hormones and Adaptation to the Environment—I. Chromaffine Tissue and Stress

Chromaffine Tissue

The adrenal gland consists of two components, an outer cortex and an inner medulla, distinct in their structure, functions and their embryological origin. The medulla forms the greyish core of the gland, containing groups or cords of cells surrounded by blood vessels. In contrast to the cortical zone the medulla stores large amounts of its hormones, the catecholamines adrenaline and noradrenaline. These hormones because of their component phenolic and alcoholic groups are readily oxidized (fig. 3). Condensation of the resulting oxidation products produces self-coloured polymers. This is the chemical basis for the chromaffine reaction of the adrenal medulla, the darkening of the tissue after exposure to solutions containing chromic acid or bichromate. This reaction was observed early in the history of the study of the adrenal medulla. It was originally thought that medullary tissues had a specific affinity for chromium, hence the term chromaffine tissue. However, we now know that the reaction is not specific to adrenaline or noradrenaline but is shown by a whole series of aromatic reducing substances, which include biological amines such as dopamine and 5-hydroxytryptamine. Furthermore, the reaction can be obtained using oxidizing agents other than chromium compounds. Therefore the term chromaffine tissue is somewhat inappropriate, but it is retained for historical reasons and because of its extensive usage.

Chromaffine tissue is not restricted to the adrenal medulla but is widely distributed in the mammalian body from the base of the skull to the pelvis. Both medullary and extra-medullary chromaffine tissues produce the hormones adrenaline and noradrenaline. In fact these tissues have an identical origin in the embryo, arising as neuroblasts (primitive nerve cells) mainly in the thoracic part of the central nervous system. These primitive nerve cells migrate from their place of origin along the course of the sympathetic nerves. Some of these

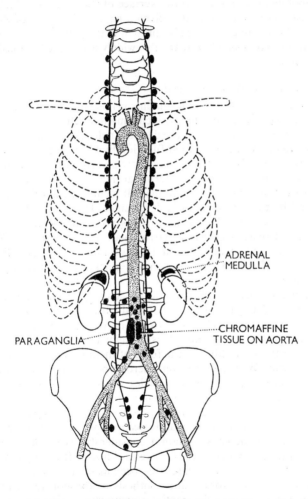

FIGURE 25. Diagram showing the distribution of chromaffine tissue, both adreno-medullary and extra-medullary. Extra-medullary tissue is shown along the sympthatic nerve chain (ganglia not shown) and related to the blood vessels in the abdomen and pelvis.

cells which migrate to the dorsal abdominal wall become associated with adrenocortical tissue to form the adrenal medulla. The extra-adrenal chromaffine tissue is widely distributed. Some of it is aggregated to form discrete, easily recognized bodies particularly in the abdomen where they form the paraganglia in or near the ganglia of the sympathetic nerve chains. The aggregation of the extra medullary chromaffine tissue on the ventral surface of the aorta is so large and so consistently placed that it is given a name—the organ of Zuckerkandl. Chromaffine cells are also present in the pelvic nervous plexus, in relation to the gonads and are also widely distributed in the dermis of the skin.

Structure of the chromaffine cell

The structure of a typical chromaffine cell is shown in fig. 26; nerve fibres are typically associated with chromaffine cells. These endocrine cells which secrete hormones are derived in development from neuroblasts and are regulated in their function by fibres of the sympathetic nervous system. Nerve fibres may show a variable relationship with the chromaffine cell. The nerve terminals may be at the surface of the cell or the fibre may invaginate the cell membrane and become surrounded by the cell for a variable distance. However, whatever the form of the relationship between the nerve terminal and the chromaffine cell there is always a narrow gap, only about 150 Å wide between the cell membranes of the two units. The outer limiting membrane of the two components is thickened in the region of contiguity.

The nerve fibres which regulate the activity of chromaffine cells exert their effect by the liberation of small, discretely placed quantities of the activator acetylcholine into the narrow gap between the cell membranes of the two units. It is thought that the acetylcholine is present in the nerve fibre in the form of membrane-bound stores of the activator. With the electron microscope it is possible to detect these vesicles called synaptic vesicles measuring about 350 Å diameter. The release of acetylcholine from these stores into the synaptic cleft causes the release of the chromaffine cell hormone, either noradrenaline or adrenaline according to the cell type.

The chromaffine cells contain a considerable store of hormones, sufficient to satisfy body needs for several hours. It is only in the last 10 years or so that some progress has been made in the understanding of the methods of storage and synthesis of the hormones in these cells. Early work using differential centrifugation of homogenates of the adrenal medullary cells localized the hormone to the so-called mitochondrial fraction. However, later it proved possible to separate stored hormone from the mitochondria by suspending the mitochondrial fraction in isotonic sucrose solution and centrifuging it in a particular gravity gradient. At about the same time electron micrographs of chromaffine cells demonstrated the presence of many large granules related to the endoplasmic reticulum of the cell.

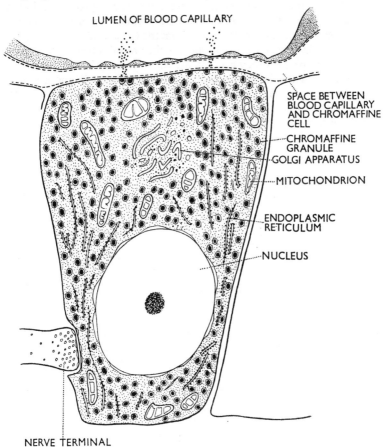

FIGURE 26. Diagram of the chromaffine cell.

Electron microscope studies of the hormone-containing granules obtained by the density gradient centrifugation of the chromaffine cell homogenates leaves little doubt that these are identical with the dense granules seen in the electron micrographs of intact chromaffine cells. There is also evidence that some of these granules contain noradrenaline, while others contain adrenaline. One striking feature of the granules is that they contain large amounts of adenosine triphosphate (ATP) with some adenosine diphosphate (ADP) and adenylic acid (AMP). It is suggested that the phosphate groups of ATP serve to neutralize basic groups on the hormones. These granules containing ATP-bound hormone are related to the endoplasmic reticulum of the cell. The granules are composed of protein and lipid and are surrounded by an external membrane. It appears that the dense centres of the granules arises from the Golgi apparatus and may consist of binding substance alone. As the granules move away from the Golgi apparatus they accumulate the hormone. In the process of the release of the hormone from the chromaffine cell, the granules appear to approach the cell membranes where there is a fusion of the outer membrane of the granule with the cell membrane. The contents of the granule are discharged from the cell by a reverse process of pinocytosis, whereby they disintegrate and liberate their contained hormone and ATP. Intact granules are not normally seen in the circulating blood.

It has been suggested that the following sequence of events occurs on stimulation of the splanchnic nerve:

1. The release of acetyl choline from sympathetic nerve terminals and the diffusion of acetylcholine across the narrow synaptic space to the cell membrane of the chromaffine cell.
2. Acetylcholine reacts with specific sites, called receptors, on the chromaffine cell and this triggers off various reactions. Observations on isolated chromaffine cells *in vitro* show that there is a depolarization of the membrane. There is also an increased ionic permeability of the membrane.
3. Calcium ions migrate across the activated membrane into the interior of the cell and arrive at certain critical sites. The rôle of the calcium ion in excitation–secretion coupling has been demonstrated in various ways, e.g. adrenal medullary cells do not respond to acetylcholine when calcium is removed from the perfusing fluid.
4. This process in some unknown way causes the release of catecholamines and ATP from the storage granules and these substances gain access to tissue fluid and blood. After stimulation of the splanchnic nerve there is a prompt increase in the concentration of adrenaline, noradrenaline and ATP in adrenal venous blood.

This mechanism of activation shows close similarity to the process of neuromuscular activation.

Differences between adrenal and extra-adrenal chromaffine tissue

Extracts from the adrenal medulla contain two hormones, noradrenaline and its methylated derivative adrenaline. The proportion of these two catecholamines varies from species to species and during the development of a particular species. In most mammals noradrenaline is the predominant amine of the adrenal medulla before birth and in early neonatal life, but adrenaline eventually becomes the more important hormone quantitatively as the animal gets older. The rate at which this change occurs varies in different species and in man this is not achieved until the age of three years. However, extraadrenal chromaffine tissue continues to secrete a preponderance of noradrenaline in adult life. In this it resembles the post-ganglionic neurones of the sympathetic nervous system.

The chromaffine cells which are embedded in the adrenal gland therefore are peculiar in that they produce adrenaline rather than noradrenaline during adult life, and this poses the interesting question of the significance of the relationship of the two tissue components of the adrenal gland. This relationship of the cortical and medullary tissue is phylogenetically fairly recent in origin in that in fishes and amphibia the equivalent of the adrenocortical tissue is separate from the chromaffine tissue. It has been suggested that the high concentration of the adrenocortical hormones in the venous blood which leaves the cortex and drains into the medulla on its way into the systemic circulation is somehow responsible for the production of adrenaline from noradrenaline. Perhaps this high concentration of hormones induces the synthesis of the appropriate methylating enzyme which converts noradrenaline to adrenaline. Adrenaline-producing cells are arranged in palisade form surrounding thin-walled blood vessels which usually serve to drain blood from the cortical region to the central vein of the adrenal gland. The blood vessels adjacent to the noradrenaline-secreting cells are more associated with the arterial blood supply to the medulla, that is to blood containing a lower concentration of adrenocortical hormones. These two cell populations of the adrenal medulla, the adrenaline secreting and the noradrenaline secreting, have been identified by a variety of staining procedures which distinguish the two hormones.

In some respects adrenaline is a more potent activator than noradrenaline and this may be of adaptive significance in the development of this complex relationship. The two hormones have rather similar properties but their threshold of action differs considerably. Comparing the activity of equal molecular doses, adrenaline is in general four to eight times more potent than noradrenaline. There are some qualitative differences in the effects of the two hormones, which may be of adaptive significance in the evolution of the adrenal medulla as a secretor of adrenaline rather than noradrenaline. For example, noradrenaline has widespread vasoconstrictor effects which causes an increase in total peripheral resistance and thus blood pressure. Adrenaline is also a potent vasoconstrictor, but at the same time it has vasodilator effects in some vascular beds. Thus adrenaline can alter the pattern of blood flow in favour of certain organs on which it has a vasodilator effect. This effect may be of significance, for example, in cardiovascular adaptation to cold exposure or to muscular exercise. Further, noradrenaline has virtually no effect on glycogen metabolism so that, in contrast to adrenaline, it cannot participate to any degree in the regulation of the glucose content of the blood.

The physiological effects of adrenaline and noradrenaline

In general the effects of the administration of adrenaline and noradrenaline are similar to those produced by the stimulation of the sympathetic nervous system. In fact the liberation of the hormones into the blood occurs coincidently with the activation of the sympathetic nervous system, and the hormones reinforce and elaborate the action of the sympathetic nervous system. There is an enormous literature on the effects of these two hormones on a variety of organs and systems so that only a very brief summary of these effects can be given here. This summary will consider the rôle of chromaffine tissue in the whole organism.

CARDIOVASCULAR SYSTEM

The hormones produce widespread and complex actions on the cardiovascular system. The effect of intravenous injection of small amounts of adrenaline or noradrenaline is dramatic. The heart rate is quickened, the vigour of the contraction of the heart is increased and these two effects result in an increase in the volume of blood leaving the heart. At the same time as these cardiac effects occur there is a vasoconstriction in some arteriolar beds. Thus there is an increase in the volume of blood in the large arteries and the pressure of the blood in these arteries rises. Adrenaline, although it has potent effects on the heart, produces less effect on blood pressure than does noradrenaline because adrenaline has a vasodilator effect on certain arteriolar beds. Noradrenaline produces a widespread vasoconstriction, involving the blood supply to the skin, brain, kidney, abdominal viscera and muscles. Although adrenaline produces vasoconstrictor effects on the vessels supplying the skin and kidney, it produces vasodilation of vessels supplying the brain and muscles. Thus adrenaline redistributes the blood and this is of adaptive significance, for example, during muscular exercise.

These substances have such potent effects on the cardiovascular system that the injection of the hormones for therapeutic purposes in man is associated with considerable hazards. The effects of the muscle of the heart are associated with an increased excitability of the heart;

there is a shortening of the refractory period of the auricular muscle, an increase in the speed of conduction of the cardiac impulse and an increase in the automaticity of the heart muscle. These changes may precipitate abnormal cardiac rhythms which can be lethal if the ventricle becomes involved.

METABOLIC EFFECTS

A prominent effect of catecholamines is on carbohydrate metabolism. This is illustrated by the early rise in the concentration of glucose in the blood following the administration of adrenaline. There is also a rise in the lactic acid concentration of the blood. Associated with these effects is a fall in the glycogen content of the liver and muscles. The mode of action of adrenaline on glycogen breakdown in the liver is discussed in some detail in chapter 3. In striated muscle adrenaline acts in a similar fashion on the hydrolysis

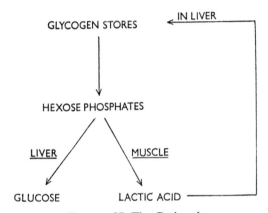

FIGURE 27. The Cori cycle.

of glycogen stores by the activation of the enzyme phosphorylase and an early observed effect of adrenaline is an increase in the concentration of hexose phosphates in muscle. Muscle lacks the enzyme glucose-6-phosphatase (which is possessed by the liver) and instead of acting as a source of blood glucose the hexose phosphates produced from glycogen enter the metabolic pathways in the muscle. The increased metabolism of hexose-phosphates in muscle is associated with the production of lactic acid. Much of this passes out of muscle cells into the circulating blood and much of the lactic acid is metabolized by the liver with the production of glucose and glycogen. This sequence of events known as the Cori cycle is summarized in fig. 27. Adrenaline is very effective in producing these effects on carbohydrate metabolism and does so in doses less than those needed to raise the blood pressure. Noradrenaline, however, has minimal effects on carbohydrate metabolism.

Catecholamines also have effects on fat metabolism and the administration or release of the hormone causes a mobilization of fat reserves, an increase in the metabolic rate of adipose tissue and a rise in the concentration of fatty acids in the circulating blood. The significance of these effects of the hormone in temperature regulation is discussed in chapter 11.

SMOOTH MUSCLE

In addition to the effects of catecholamines on the smooth muscle of the vascular system there are widespread areas of smooth muscle which are sensitive to these hormones. The hormones induce contraction of the pilomotor muscles (causing erection of the hairs), relaxation of the bronchial musculature (reducing the resistance to the passage of air into the lungs), relaxation of the intestine, contraction of the ureter and vas deferens and contraction of the dilator muscles of the iris. The response of the uterus to adrenaline and noradrenaline vary from species to species and according to the stage of the oestrous cycle or pregnancy.

STRIATED MUSCLE

By contrast to the effects of the hormones on smooth muscle and cardiac muscle, catecholamines do not induce direct stimulatory or inhibitory effects on striated muscle. They do, however, modify the responses of striated muscle to stimulation by way of its motor nerves. As first shown by Oliver and Schafer in 1895, adrenaline prolongs the mechanical response of striated muscle during tetanic stimulation and it also increases the response of striated muscle after partial fatigue has developed. These effects were initially considered to be due to some influence on neuromuscular transmission. However, it is now known to occur whether the muscle is stimulated via its nerve supply or via electrodes placed directly on the muscle. It is thought that this effect of adrenaline is explicable in terms of its effect upon carbohydrate metabolism of muscle. During the initial stages of muscular contraction the hydrolysis of glycogen under the influence of phosphorylase provides the glucose needed for oxidation in order to provide the ATP requirements of contraction. However, phosphorylase is inactivated during muscular activity. By promoting the regeneration of active phosphorylase, adrenaline can assist in the provision of continued supplies of glucose and hence ATP and so can permit the muscle to continue to respond to stimulation.

NERVOUS SYSTEM

A variety of nervous effects follow the intravenous injection of catecholamines or their release within the organism. In man the administration of adrenaline may produce restlessness, tremor and apprehension. In experimental animals injection of adrenaline into the carotid artery causes excitement, exaggerated muscular responses and spasm of muscles, which may progress to convulsions. The study of the effects of catechol amines on the nervous system, in particular the effects on higher nervous centres, is still in its early stages. A large number of different effects of administered adrenaline have been reported, including marked cortical arousal, facilitation of postural reflexes, effects on descending facilitatory systems in the reticular formation, effects on evoked cortical potentials, inhibitory effects via baroreceptor reflexes, effects on sympathetic ganglionic transmission and neuro-endocrine effects, such as the activation of the release of ACTH from the anterior pituitary gland. Some of these effects will be briefly considered.

(i) *Cortical arousal.* When the animal is aroused from light sleep by means of a non-specific stimulus the electrocorticogram becomes 'desynchronized', that is a high amplitude, slow sinusoidal activity becomes replaced by a low amplitude fast activity. The effect is mediated by way of special pathways in the brain stem—the reticular activating system. The cortical arousal and facilitation of postural reflexes which follows the administration of adrenaline is due to an effect of the hormone on the rostral area of the descending and ascending brain stem reticular formation. When this part of the reticular core is destroyed, adrenaline can no longer produce its effects on cortical arousal.

(ii) *Effects on the pituitary gland.* Adrenaline has been shown to increase the secretion of various hormones from the anterior pituitary gland, including thyrotrophic hormone, gonadotrophic hormone and ACTH. The real significance of these effects is not fully understood. However, some effects on the organism previously attributed to adrenaline can now be shown to be due to the release of pituitary hormones. Thus the dissolution of lymphatic tissue, lymphopenia and eosinopenia, which were formerly regarded as the effect of adrenaline can now be attributed to ACTH and activation of the adrenal cortex. Very small amounts of adrenaline will activate the pituitary-adrenal mechanism. Since adrenaline is released under the same conditions of stress which activate the adrenal cortex it is possible that adrenaline is one of the factors in triggering adrenocortical activity during stress. However, since animals from which the adrenal medullary tissue has been removed do not suffer from adrenocortical deficiency during stress, then adrenaline cannot be the only factor involved in adrenocortical regulation.

The regulatory effect of adrenaline on ganglionic transmission and neuromedullary transmission is discussed on p. 114.

The regulation of the secretory activity of chromaffine tissue

The secretory activity of chromaffine tissue is determined by its connections with the sympathetic nervous system. Section of the nerves reduces the hormone output to a low level, which remains constant and unaffected by physiological requirements. The adrenal medulla and extramedullary chromaffine tissue are equivalent on a functional level to sympathetic ganglia; their secretory cholinergic fibres correspond to the preganglionic sympathetic nerve fibres and the chromaffine cell itself corresponds to the postganglionic nerve fibre. The organization of nerve fibres regulating the adrenal medulla is similar to that of the sympathetic vasoconstrictor system. The higher centres controlling the adrenal medulla are located in the cerebral cortex, hypothalamus and medulla. The adrenal secretory pathway is projected from these higher centres down the spinal cord and emerges in the thoracic and upper lumbar regions.

Factors activating chromaffine tissue

All forms of stress which produce a general activation of the sympathetic nervous system also activate chromaffine tissue. Thus chromaffine tissue acts in conjunction with the vasomotor system in the regulation of the arterial blood pressure. During a fall of blood pressure, e.g. induced by haemorrhage, there is a widespread activation of the sympathetic nervous system and a release of adrenaline and noradrenaline from chromaffine tissues. Many other forms of stress activate chromaffine tissue, including hypoglycaemia, cold, asphyxia, muscular exercise and emotions.

Some factors can reduce the secretory activity of the chromaffine tissue. A rise of arterial blood pressure, e.g. due to administration of noradrenaline, will automatically curb the endogenous release of the hormones, acting by way of the sino-aortic baroreceptors which discharge inhibitory impulses to the central nervous system, which ultimately results in a reduced amount of activation of the chromaffine tissue. Adrenaline can also reduce chromaffine cell secretion, even without the development of hypertension. Thus perfusion of the carotid sinus of an animal under constant hydrostatic pressure with a solution containing adrenaline, will reduce adrenomedullary secretion, and this action is regarded as a direct effect of the hormone on the muscle of the carotid sinus wall. A central braking effect of adrenaline has also been described and is seen in animal preparations in which the head of an animal is perfused under constant pressure with solutions of adrenaline after the elimination of all nervous reflex mechanisms involved in vasomotor control. Noradrenaline can also exert a similar braking action on the secretory activity of the adrenal medulla.

The physiological significance of chromaffine tissue

The function of the adrenal medulla has been regarded by Cannon and others as an emergency mechanism which is activated during periods of stress, thereby preparing the animal for 'fight or flight'. The actions of adrenaline and noradrenaline on cardiovascular function, glucose metabolism, striated and smooth muscle function and on nervous activity are of adaptive significance.

However, an alternative view is that the secretion of these hormones is a continuous process with periods of reinforcement occurring during stress. In this view the hormones would be involved in the various homeostatic mechanisms, e.g. regulation of blood pressure, blood glucose, etc. It has proved difficult to demonstrate what are in fact the basal levels of secretion of adrenaline and noradrenaline from the adrenal medulla, since the various experimental conditions at the time of the sampling of blood for analysis can affect the results. The process of catheterizing the adrenal vein for sampling—and only such samples truly determine medullary activity—can increase the secretion rate. Even if blood loss is minimal with no change in arterial blood pressure, the secretory activity of the medulla can increase threefold. These factors account for variations in measured adrenaline and noradrenaline content of adrenal venous blood. When precautions are taken to minimize the interfering factors, then on average the adrenal glands secrete 0·04 μg/kg/min of adrenaline and 0·05 μg/kg/min of noradrenaline. These values were obtained by workers using a non-anaesthetized animal and are close to the level of the 'paralytic secretion' which is found after section of the nerves to the adrenal glands. These findings support the view that secretion from the glands is an 'emergency mechanism'.

One can regard the adrenal medulla and extra-medullary chromaffine tissue as a specialized component of the sympathetic nervous system which produces in large quantities the hormones which are liberated in much smaller amounts at sympathetic postganglionic terminals. The hormone adrenaline, released from the adrenal gland is of special significance, since it is a much more potent substance than noradrenaline, which is the predominant mediator at the terminals of the sympathetic nervous system. In the words of Tournade 'the splanchnic nerve delegates to an internal secretion the power to start various organic activities which it directly controls itself. It does it so well that by this very economical process its action is reinforced and gains in extent and duration'. The reinforcement to which Tournade refers is of a very direct nature for not only does adrenaline mimic most of the actions of sympathetic nerve stimulation but it also has a selective action in potentiating sympathetic ganglionic and neuro-adrenomedullary transmission. Small amounts of adrenaline applied to sympathetic ganglia reinforce synaptic transmission in the ganglia so that stimulation of the preganglionic nerves produces effects which are more prolonged and more pronounced. Similarly, adrenaline can enhance the effect of acetylcholine on the chromaffine cells. Thus it can be seen that adrenaline intensifies the action of the sympathetic nervous system. Given the same level of excitation of the central sympathetic centres, adrenaline will intensify the peripheral effects of sympathetic discharge.

Early investigations of the rôle of the medulla in the organism exaggerated the significance of the medulla since the investigations were made before the importance of the adrenal cortex was recognized. Later studies in which demedullation of the adrenal glands was made without serious damage to the cortices showed that demedullation is well tolerated and consistent with life. Adrenaline disappears from the urine of man and dogs after demedullation. Noradrenaline levels, if they are changed at all, are increased, probably due to increased activity of the extra-medullary chromaffine tissue. It would seem then that the adrenal medulla is not essential for life. However, the significance of this apparent lack of effect of adrenal demedullation has sometimes been questioned because of the widespread distribution of extra-medullary chromaffine tissue, which could take over to some degree the function of the adrenal medulla. Indeed, increased activity of the sympathetic nervous system itself might obscure the effects of demedullation.

Mode of action of catecholamines

The way in which adrenaline produces an increased rate of production of glucose from glycogen in the liver is described in detail in chapter 3. To summarize the events which occur in the liver following the administration of adrenaline, we can say that the hormone interacts with the adenyl-cyclase system which is located in or near the cell membrane, to cause an increased rate of formation of the substance cyclic 3′,5′-AMP. This nucleotide causes an increase in the amount of active phosphorylase. The activated phosphorylase leads in turn to an increased rate of glycogen breakdown.

Catecholamines have been found to produce similar effects on cyclic, 3′,5′-AMP formation in a variety of other tissues, including skeletal muscle, heart, adipose tissue, uterus, intestinal smooth muscle, spleen, lung, brain and frog skin. In skeletal muscle cyclic 3′,5′-AMP has an effect on the enzyme phosphorylase similar to its action on liver phosphorylase. However, because of differences in the enzyme composition of liver and muscle cells, in particular the lack of glucose-6-phosphatase, the effect of the hormone on muscle tissue is somewhat different. Glycogen stores become depleted because of the activity of phosphorylase, but the glucose-6-phosphate produced is not transformed into glucose. Instead there is an increased rate of metabolism of glucose-6-phosphate, resulting in lactic acid formation which passes out of the muscle into the circulating blood.

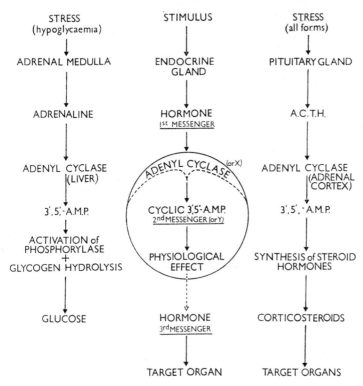

FIGURE 28. Second and third messenger concept. The central scheme shows the basic concept. On the left is an example of a two-messenger system (adrenaline and 3′,5′-A.M.P.). On the right is an example of a three-messenger system (A.C.T.H., 3′,5′-A.M.P. and corticosteroids).

The action of cyclic 3′,5′-AMP is not restricted to the enzyme phosphorylase. A large number of metabolic processes has been found to be influenced by the nucleotide, including sugar transport by the thyroid gland, water permeability of the toad bladder, the conversion of lactate to glucose, synthesis of liver proteins, ketogenesis, steroidgenesis and the activity of enzymes such as phosphofructo-kinase, lipase and tryptophan pyrolase. Moreover, the accumulation of 3′,5′-AMP in tissues is not specifically influenced by catechola-mines. Many other hormones seem to exert their effects in a similar way. The table below shows some other hormones which also cause an accumulation of 3′,5′-AMP in tissues.

Hormone	Tissue
Histamine	Gastric mucosa
Acetylcholine	Heart
5-Hydroxytryptamine	Fasciola hepatica (liver fluke)
Glucagon	Liver, adipose tissue
ACTH	Adrenal cortex, adipose tissue
TSH	Thyroid
LH	Corpus luteum
Vasopressin	Toad bladder, frog skin, kidney, adipose tissue

Hormones which produce an accumulation of 3′,5′-AMP in tissues

These observations have given rise to the concept that some hormones act by two or more messenger systems. According to this concept the first messenger is the hormone released from a particular gland following appropriate stimulation. This first messenger is trans-ported in the blood stream to its target organ, where it causes the production of the second messenger. The only second messenger known at present is cyclic 3′,5′-AMP. Variations in the molecular configuration of adenyl cyclase—the 'receptor' for the first messenger —from tissue to tissue, together with tissue variations in the structure of other enzymes or in the presence of other enzymes, could account for variations in the responsiveness of tissues to different hormones. Thus adrenaline produces different effects in the liver and muscle although there is a common mediator (second messenger) in the two tissues. Differences in the response of the two tissues is due to the lack of the enzyme glucose-6-phosphatase in skeletal muscle. A third messenger is known to be present in those tissues which respond to the action of the second messenger by the production of another hormone (third messenger) which leaves the organ to have an action on other tissues. This concept is illustrated in fig. 28.

10

Hormones and Adaptation to the Environment—II. The Adrenal Cortex and Stress

---

The importance of the adrenal cortex for life

The adrenal cortex is an endocrine organ which is indispensable for normal life. When the glands are removed from an animal, or are subject to disease, a whole range of defects appears, which ultimately prove fatal. If the operation of adrenalectomy is performed without too much shock, then the animal recovers from the anaesthetic and may appear normal. Sooner or later, usually in a matter of weeks, there is a general deterioration in the condition of the animal, with the appearance of various signs and symptoms, including loss of appetite, vomiting, diarrhoea, a fall in blood pressure, a failure of renal function, loss in weight, failure of growth and hypoglycaemia which may progress to convulsions. Death may occur earlier than anticipated if the animal is exposed to 'stress' in any form, e.g. injury, exposure to heat or cold, infection, starvation, etc. Some of the defects of adrenalectomized animals can be understood by considering the established physiological effects of adrenocortical hormones. The known influences of these hormones include:

1. Carbohydrate, protein and lipid metabolism.
2. Electrolyte and water metabolism.
3. Renal function.
4. Maintenance of blood pressure.
5. Effects on the central nervous system.
6. Rôle in stress, e.g. infection, injury, extremes of temperature.
7. Effects on inflammatory responses of tissues.

Some of these activities are dealt with in some detail in other chapters (carbohydrates and protein metabolism in chapter 3, electrolyte and water metabolism in chapter 7, temperature regulation in chapter 11).

Some of the defects in electrolyte balance and water balance that appear after the removal of the adrenal glands can be corrected by permitting the animal to drink salt solution or by administering aldosterone or synthetic mineralocorticoids. This aspect of adrenal function will not be dealt with further in this chapter. Here we are concerned with a more nebulous function of the adrenal cortex—the ability to confer resistance to stress, which is mediated by gluco-corticoid hormones. Some forms of resistance to stress can be readily explained by reference to known effects of glucocorticoid hormones. Thus during starvation glucocorticoid hormones from the adrenal cortex assist in the mechanisms for maintaining the level of glucose in the blood, by promoting an increased rate of gluconeogenesis from protein. In the absence of the adrenal cortex fasting is badly tolerated and may produce a fatal hypoglycaemia.

Cortical hormones confer resistance to other types of stress, e.g. cold, heat, severe muscular exercise, infection, physical injury, but here the fundamental mechanism by which corticosteroids act is unknown. Part of this protection from the effects of stress is probably due to the effect of cortical hormones on the internal distribution of salt and water between cells and tissue fluids, and the interactions of corticoids with the cardiovascular system which permit the cardio-vascular system to adapt to the changing external and internal en-vironment, e.g. the haemodynamic responses to cold exposure or to muscular exercise. There are specific interactions of corticoids with the autonomic nervous system and the cardiovascular system. Adrenocortical deficiency may result in disorders of the myocardium, a fall in blood pressure and changes in the permeability of the capillaries. In the absence of cortical hormones there are deficient responses of the cardiovascular system to noradrenaline. Repeated injections of noradrenaline into the adrenalectomized animals results in a gradual failure of responses of the blood vessels to the sympa-thetic transmitter substance. These responses can be restored by the injection of adrenocortical hormones. However, no single site of action can explain the rôle of adrenocortical hormones in providing resistance to the various forms of stress.

The mode of action of stress on the adrenal cortex

The manifold influences—called stressors by Selye—which can acti-vate the adrenal cortex, suggest that they act by a common mechanism. One component of this mechanism is the anterior pituitary gland. The activity of the adrenal cortex is almost completely regulated by the anterior pituitary by way of adrenocorticotrophic hormone (ACTH). Removal of the anterior pituitary gland results in all the defects which are produced by adrenalectomy, except that salt and water metabolism can still be regulated in a fairly normal fashion. This is because aldosterone secretion continues after hypophysectomy. Analysis of adrenal venous blood in dogs and rats has shown that although the output of glucocorticoid hormones falls to low levels, the secretion of aldosterone continues at about half the normal rate. The removal of the anterior pituitary gland also prevents the activation of the adrenal cortex by stress.

The pituitary-adrenal messenger, ACTH

Pituitary adrenocorticotrophic hormone is a polypeptide with a molecular weight of about 42,000 and it has been recently synthesized (1961). The administration of this hormone to an animal results in an increase in adrenal weight and an increased rate of synthesis and release of cortical hormones into the blood. If the ACTH is administered to a previously hypophysectomized animal the actions on the adrenal cortex results in an increased resistance to stress. The response of the adrenal cortex occurs within minutes of the administration of the hormone.

Control of the release of ACTH

The various stressors, e.g. heat, cold, light, atmospheric pressure, fear, pain, frustration, infections, chemical poisons and the like, exert, via nervous pathways, a common effect on the hypothalamus. This common effect is the liberation of a chemical mediator—corticotrophin releasing factor—(CRF) into the hypothalamico-hypophyseal portal circulation. The hypothalamic control of anterior pituitary function is the subject of a separate chapter (p. 140). Here we can say that if the pituitary gland is deprived of hypothalamic influences, either by adequate section of the pituitary stalk or transplantation of the pituitary away from its normal connections, then the pituitary gland no longer responds to stress by the liberation of increased quantities of ACTH into the circulating blood.

There is some evidence that suggests that the nervous system may control the pituitary secretion of ACTH by a dual mechanism, at least one excitatory (CRF), and another inhibitory. Egdahl (1961) reported that the removal of all brain tissue down to the level of the pons, leaving the pituitary isolated in its bony socket, resulted in an increased secretion of glucocorticoid hormones from the adrenal gland and this persisted during the period that the animals survived

(2–4 days). Egdahl put forward the idea that these observations were consistent with the view that the brain secreted some inhibitory factor which depresses the secretion of ACTH by the pituitary. He also found that the administration of a barbiturate depressed the activity of the adrenal cortex in these animals which had been operated upon. Stimulation of peripheral nerves, e.g. sciatic nerve, restored the secretion of the adrenal cortex. He concluded that nervous stimulation resulted in the release of a substance from the hind brain which stimulated the release of ACTH. The relationship between this factor —Egdahl's hind-brain factor—and CRF is not established.

FEEDBACK MECHANISMS
The administration of glucocorticoid hormones to an animal results in a depression of pituitary ACTH secretion which is inevitably followed by a depression of endogenous adrenocortical secretion. It is not certain at what level glucocorticoids inhibit pituitary function, i.e. whether the hormone acts on the pituitary gland or on the release of CRF from the hypothalamus.

It was suggested that this effect of corticoids could play a part in the responses of the animal to stress. It was thought that stress might increase the 'utilization' of adrenocortical hormones by various tissues, resulting in a lowering of the level of corticoids in the blood. This effect would lessen the inhibitory effect of corticoids on the secretion of ACTH by the pituitary resulting in an outpouring of ACTH and an activation of the adrenal cortex. However, direct measurements of blood levels of corticoids during stress show no such initial fall in their concentration. Furthermore, the response of the pituitary gland to stress is so rapid that it seems unlikely that such a mechanism could operate. However, it is possible that the effect of cortical hormones in suppressing ACTH release from the pituitary may operate to limit the duration of the response of the pituitary gland to stress.

One more clearly defined rôle of the adrenal gland in regulating the pituitary secretion of ACTH during stress, is the secretion of adrenaline from the adrenal medulla. The sympathetic nervous system is activated during all forms of stress (chapter 9) and part of this effect is an outpouring of adrenaline from the adrenal medulla into the circulating blood. Adrenaline is a potent activator of ACTH secretion from the pituitary gland and reinforces the other mechanisms which activate the pituitary gland during stress. In this sense the adrenal medullary and adrenocortical responses to stress are synergistic. The activity of the adrenal medulla initiates those physiological responses which adapt the animal to 'fight or flight' (p. 113) and is also involved in the activation of a second mechanism which supports the animal to resist the stress incurred in 'fight or flight'. Thus the

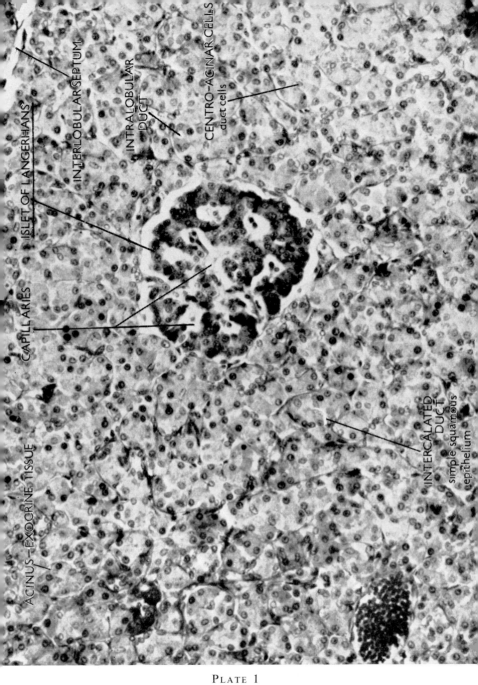

CAPILLARIES

ISLET OF LANGERHANS

INTERLOBULAR SEPTUM

INTRALOBULAR DUCT

CENTRO-ACINAR CELLS
duct cells

ACINUS—EXOCRINE TISSUE

INTERCALATED DUCT
simple squamous epithelium

PLATE 1

Photomicrograph of section of pancreas showing the islets of Langerhans. (Mag. approx. 400 ×). From Freeman, W. H. and Bracegirdle, B. B., *An Atlas of Histology.* Heinemann Educational Books Ltd., 1966.

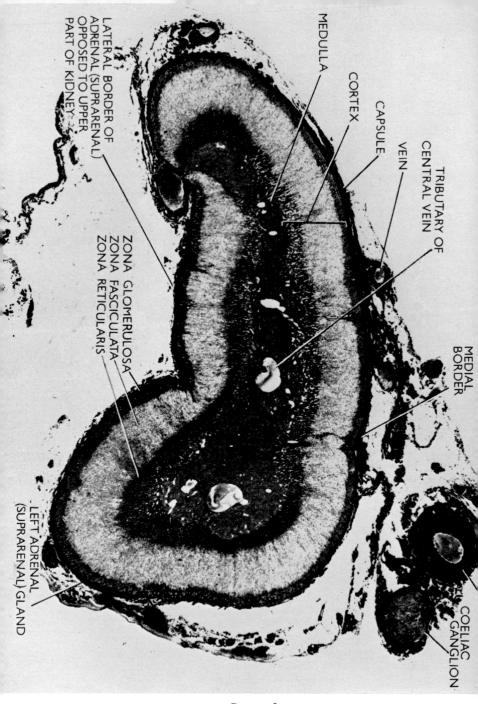

LATERAL BORDER OF ADRENAL (SUPRARENAL) OPPOSED TO UPPER PART OF KIDNEY

MEDULLA

CORTEX

CAPSULE

VEIN

TRIBUTARY OF CENTRAL VEIN

ZONA GLOMERULOSA
ZONA FASCICULATA
ZONA RETICULARIS

MEDIAL BORDER

LEFT ADRENAL (SUPRARENAL) GLAND

COELIAC GANGLION

PLATE 2
L.S. adrenal gland of monkey. (Mag. approx. 20 ×).
From Freeman, W. H. and Bracegirdle, B. B., *An Atlas of Histology*.
Heinemann Educational Books Ltd., 1966.

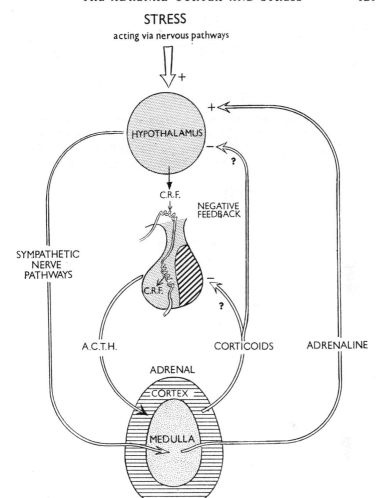

STRESS

acting via nervous pathways

FIGURE 29. Mechanisms activating the adrenal cortex in stress.

effects of cortical hormones provide the increased supplies of glucose which may be needed in fight or flight, and confers resistance to the effects of injury, infection or blood loss which may be incurred. All along the line of the transfer of the initial message from the hypothalamus to the adrenal cortex there is amplification of the message. Thus $0 \cdot 1 \ \mu g$ of a purified preparation of corticotrophin releasing factor is estimated to release $1 \cdot 0 \ \mu g$ ACTH from the anterior pituitary

E

gland. 1·0 μg of purified ACTH causes the release of 40 μg of corticoids from the isolated adrenal cortex of the rat, and this quantity of cortical hormones causes the deposition of 40 mg of glycogen in the liver of the rat.

Mode of action of ACTH

Although pituitary ACTH causes a stimulation of protein synthesis in adrenocortical tissue and causes an increase in the rate of synthesis of RNA the effects of the hormone on the synthesis of steroids by slices of adrenocortical tissue *in vitro* is independent of RNA syn-

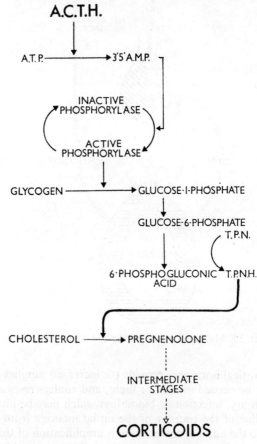

FIGURE 30. Scheme showing a suggested mode of action of A.C.T.H. on the adrenal cortex.

thesis and is not blocked by the antibiotic actinomycin D (p. 43). These findings indicate that ACTH does not produce its effects on steroid hormone synthesis by an action at the level of the gene. One clear effect of ACTH on adrenocortical tissue is to cause an increased conversion of adenosine triphosphate (ATP) to 3',5'-adenosine monophosphate cyclic ester (3',5'-AMP). This effect on the production of 3',5'-AMP follows the application of various other hormones to their target tissues. Indeed, the concept has arisen that 3',5'-AMP acts as a 'second messenger' in the target tissues (p. 115).

The accumulation of 3',5'-AMP in the adrenal cortex following treatment with ACTH causes an activation of the enzyme phosphorylase and this effect is associated with an increased rate of synthesis of the steroid hormones. It has been found that the addition of 3',5'-AMP alone to adrenocortical tissue increases the rate of production of steroid hormones, just as does ACTH itself. If adrenocortical tissue which is responding to the addition of 3',5'-AMP is now treated with ACTH there is no further increase in the rate of steroid hormone synthesis which suggests that the two substances act by a common mechanism.

A possible mechanism for the stimulation of synthesis of steroids by ACTH is illustrated in fig. 30. According to this hypothesis ACTH causes an increased rate of synthesis of 3',5'-AMP which in turn activates the enzyme phosphorylase. This enzyme by accelerating the hydrolysis of glycogen, increases the supply of glucose-6-phosphate to the cell, and after the oxidation of glucose-6-phosphate it increases the provision of reduced triphosphopyridine nucleotide (TPNH). Reduced TPN is needed for the synthesis of a steroid from its precursors. This hypothesis predicts that the glycogen content of the adrenal cortex should decline after treatment with ACTH and also that the supply of various substances to the tissue, including glucose, glucose-6-phosphate and TPNH should mimic the effects of ACTH. The evidence for the effects of ACTH on glycogen content of the adrenal cortex is conflicting, but certainly the addition of both glucose-6-phosphate and triphosphopyridine nucleotide or the addition of TPNH alone will stimulate the synthesis of steroids by the adrenal cortex. Further *in vitro* experiments have shown that glucose is necessary to obtain a maximal response of the adrenal cortex to ACTH.

11

Hormones and Adaptation to the Environment—
III. Thermoregulation

Sources of body heat

The energy involved in metabolic processes in the body is ultimately derived from foodstuffs, in which it is present as the energy of chemical bonds. This energy is utilized by animals for the purpose of doing work which may be chemical synthesis, mechanical work, the establishment of electrochemical gradients, etc. Since the metabolic processes of the body are inefficient (usually less than 30% efficient), much of the energy undergoing transformation appears as heat energy. In the resting, fasting adult individual, however, all the energy of metabolic processes is ultimately transformed into heat energy. Thus although the heart may perform work in discharging blood into the aorta, the kinetic energy of the moving column of blood is ultimately converted into heat in overcoming the frictional resistance in the smaller blood vessels. Similarly, the energy involved in the chemical work performed when hydrogen ions are secreted into the stomach, appears ultimately as the heat of neutralization as stomach acid contents enter the duodenum and meet the alkaline secretion of the pancreas.

This heat, produced by metabolism, must be dissipated from the body if overheating is to be prevented. Overheating has deleterious effects, not only on the whole cell's activities (especially in neurones) but also on the properties of the molecular components of cells, e.g. the proteins, and in particular on enzyme systems. Cooling also has potent effects on cell activities. Of all the vertebrate animals only birds and mammals possess regulating mechanisms which ensure a relatively constant internal body temperature. It is this feature which enables birds and mammals to live successfully in extremes of temperature from the Arctic and Antarctic regions to the tropics. Heat-regulating mechanisms are highly complex and incompletely understood and they involve not only the regulation of heat exchange with the external environment but also the regulation of heat production by the modulation of metabolism.

It is in the modulation of the metabolic processes that the hormones such as thyroxine and noradrenaline exert their influence. However, before examining the rôle of hormones in thermoregulation we will first consider the general organization of temperature control.

Temperature-regulating mechanisms

The term 'constant body temperature' is a relative one. Body temperature even in mammals, does vary both in space and in time. Thermometers, or temperature sensitive thermocouples, applied to different regions of the body, e.g. skin, mouth, rectum, eardrum, will readily indicate that there may be marked differences in temperature of these different parts. The temperature of the outer layers of the body particularly, also show a variation in temperature at different times, as they are influenced by changes in blood flow, movement, exposure, sweating, etc. Attempts to find a direct correlation, between changes in mouth, skin, or rectal temperatures and the adaptive mechanisms for heat loss, have usually failed in the past. In 1958, however, Benzinger and his colleagues, who were studying the relations between temperature-regulating mechanisms and body temperature measurements, started to record the temperature of the tympanic membrane in the ear. This region is supplied by blood from the internal carotid artery, which supplies the brain itself. For the first time they were able to produce an experimentally reproducible relationship between stimuli and responses in heat regulation. They found that sweating began at a clearly defined value of internal cranial temperature. As the internal temperature rose, then sweating increased in an almost linear fashion. The sensitivity of this mechanism was exquisite, with measurable responses to temperature variations as small as 0·01° C. In some of these experiments the warming of the subject came from within the body, i.e. by exercise. In these cases cooling of the skin occurred because of the movement of the limbs through the cool air and because of the loss of heat from the skin in the form of latent heat of evaporation of sweat. These results imply that the temperature-regulating responses were initiated by changes in the intracranial temperature acting on a system capable of appreciating a rise in temperature of 0·01° C.

The rôle of the brain in temperature regulation was appreciated as long ago as 1884, when Aronsohn and Sachs produced hyperthermia (elevated body temperature) by damaging an area of the brain in the midline, adjacent to the corpus striatum. We now know that the area they damaged was in the hypothalamus, in the floor of the forebrain. The location of the central heat-regulating mechanism in the hypothalamus has been confirmed by many workers using techniques involving the destruction of certain parts of the hypothalamus, local heating or cooling and electrical stimulation. In the anterior region of the hypothalamus there is a region sensitive to thermal stimuli. When this region was selectively heated, in cats, by high frequency currents, typical heat loss responses were elicited, including panting and sweating. If this anterior part of the hypothalamus is damaged, for example by electrocautery, then fever develops and, if the animal survives, this is followed by a permanent defect of temperature regulation when the animal is in a hot environment. This area of the hypothalamus which protects the animal from hyperthermia, has been called the 'heat loss centre'.

The posterior region of the hypothalamus is concerned in the protection of the animal from excess cooling. Experimental lesions of the posterior hypothalamus are followed by a failure of the normal adaptive measures which should be brought into play when there is a fall in the temperature of the external environment. Following large lesions the animal becomes inactive and fails to show the normal responses of shivering and piloerection.

The hypothalamus is not only an exquisite thermostat appreciative of minute changes in deep body temperature but it also initiates those activities, such as shivering, sweating, piloerection, etc., which are aimed at returning body temperature to normal levels. In performing these tasks the hypothalamus integrates a variety of information, not only information about the temperature of blood perfusing this area but also information about the temperature of the peripheries which is transmitted from a very large number of temperature sensitive patches, particularly in the skin. There has been controversy in the past over the rôle of these two temperature-sensing systems, the central hypothalmic zone and the peripheral cutaneous zone. Both systems can, under appropriate conditions drive the thermoregulatory mechanisms. Thus during muscular exercise there is a rise of deep body temperature, due to increased metabolism of muscles, and the central system initiates heat loss responses (sweating, cutaneous vasodilatation, etc.) even although there may be a fall in skin temperature because of heat loss by movement in cooler air and in the evaporation of sweat. That the peripheral skin receptors can also play an important rôle in thermoregulation is shown by the observations of Hammel *et al.* (1963) who showed that a dog in the cold may shiver even when the temperature of the hypothalamus is normal or elevated. The drive for this regulation presumably comes from the skin receptors. The nature of the integration of information from the hypothalamus and the skin is not clear.

A possible role of chemical activators in the hypothalamus

The hypothalamus contains amounts of the biologically active substances adrenaline, noradrenaline, 5-hydroxytryptamine (5-HT), acetylcholine and histamine. According to Carlson (1962) monoamines present in the hypothalamus occur within nerve fibres which have their synapses in this region. The injection of some of the substances—notably adrenaline, noradrenaline and 5-HT—into the cerebral ventricles or into the hypothalamus itself produce marked effects on thermoregulation. These substances fall into two classes. One class of amines when injected into the cerebral ventricles or hypothalamus produces a rise in body temperature (hyperthermia) whereas injection of the second class produces a fall in body temperature (hypothermia). Thus in the cat, dog and monkey injection of adrenaline or noradrenaline produces hypothermia, whereas 5-HT produces hyperthermia. However, not all species respond similarly. The rabbit and sheep react in an opposite fashion to cats, dogs and monkeys, and respond by hyperthermia to adrenaline and noradrenaline and by hypothermia to 5-HT.

Feldberg and Myers (1964) have put forward the concept that changes in the rate of release of these amines from nerve fibres synapsing in the hypothalamus determines temperature regulation. They suggested that the amines might be continually released and that normal temperature is the outcome of a fine balance between the release of catecholamines (adrenaline and noradrenaline) and 5-HT. Changes in body temperature would be brought about by a disturbance of this balance. Alternatively, normal temperature might be maintained independently of the release of the amines, their release being the mechanism by which changes in temperature are effected. The amines are thought to act directly on the effector nerve pathways of thermoregulation, e.g. those determining shivering, patterns of blood flow, etc.

For this hypothesis it does not matter that the amines have different effects in different species of animals provided that there are two classes of amines having opposite effects in a particular species. The situation is, however, complex and Brittain and Handley (1967) found that both 5-HT and noradrenaline produced hypothermia when injected into the cerebral ventricles of the conscious mouse, whereas adrenaline caused a slight rise in temperature followed by a fall. Thus the concept of Feldberg and Myers does not appear to hold for the mouse. In man at least, temperature information from the skin arises to consciousness, causing the initiation of other adaptive measures in temperature regulation, such as seeking a cooler or warmer environment, changing the amount of clothing worn and so on.

It is necessary to stress here that the heat-regulating mechanisms do not function independently of one another, or of other body requirements. The body may have to endure a period of central hyperthermia during muscular exercise. Further, in muscular exercise, the pattern of blood flow cannot be determined solely by the needs of temperature regulation, and the blood flow to the lungs, heart and muscles increases at the expense of the cutaneous blood supply and hence at the expense of heat loss mechanisms.

Hormones in temperature regulation

Hormones act effectively in thermoregulatory processes mainly by influencing the mechanisms of heat production. Heat is produced in all living cells as a by-product of metabolism. Hormones can alter heat production of the body by causing an overall or localized increase in the metabolic rate of tissues, that is by increasing the amount of substrate utilized in unit time. In this function hormones may act by decreasing the efficiency of metabolic processes so that less of the energy contained in the substrate appears in metabolically useful forms of energy and more appears as heat. A variety of hormones are involved in thermoregulation and there is a complex interplay of one endocrine gland upon another which is incompletely understood.

Noradrenaline and thermoregulation

When adult mammals are acutely exposed to cold (i.e. are rapidly moved from a warm environment to a cold one) various processes are initiated which conserve body heat and at the same time increase heat production. Heat is conserved by a reduction of blood flow to the skin and by piloerection so as to increase the heat-insulating barrier of the skin. Heat production is increased initially by a general increase in muscular activity and by the process of shivering (i.e. rapid involuntary tremors of muscles). However, if exposure to cold continues, then various changes occur. Shivering becomes reduced in extent as other forms of heat production develop. These later forms of heat development are referred to as non-shivering thermogenesis. In addition, animals develop the ability to maintain the temperatures of the skin and the extremities above freezing point by increasing the blood flow to the skin.

NORADRENALINE AND NON-SHIVERING THERMOGENESIS

This form of increasing the production of heat is present in new-born mammals, and is indeed the principal heat-producing mechanism in the neonate. As the animal grows older, non-shivering thermogenesis is gradually replaced by shivering thermogenesis, although when the mature mammal is chronically exposed to cold the non-shivering system can be reactivated.

Non-shivering thermogenesis is regarded as an effect of noradrenaline. When mammals are exposed to cold, there are increases in the amount of circulating noradrenaline. This is supported by the fact that under these cold conditions one can observe increases in the rate of urinary excretion of noradrenaline. Rats exposed to a temperature of 3° C increased their urinary excretion of noradrenaline from resting values of 4 μg/kg/day to 20 μg/kg/day, within the first 24 hours after exposure to the cold. They maintained this high level of excretion of noradrenaline throughout the period of exposure to cold. The non-shivering thermogenesis of new-born mammals has been found to be blocked by the injection of drugs which block the transmission of nerve impulses in sympathetic ganglia, or by drugs which compete for those 'receptor sites' in the effector cells, through which noradrenaline is known to have its effects. These findings implicate the sympathetic nervous system, its chemical transmitters and the adrenal medulla (a source of adrenaline and noradrenaline) in non-shivering thermogenesis. This is supported by the finding that metabolism is stimulated by infusions of noradrenaline.

THE SITE OF NON-SHIVERING THERMOGENESIS

The first attempts to localize the tissues involved in non-shivering thermogenesis used rats, which were adapted to cold, paralysed with curare to prevent shivering and eviscerated to remove the major abdominal organs. These paralysed, eviscerated rats were capable of large metabolic responses, either to cold exposure, or to the injection of noradrenaline. An effect on muscle was shown by measurements of the amount of oxygen consumed by the muscles during cold exposure or infusion of noradrenaline. During these procedures the amount of oxygen consumed by the leg muscles doubled, without any increase in the blood flow to the muscle. In addition to increasing the metabolic rate of muscle, noradrenaline also has an effect on the metabolism of adipose tissue. Noradrenaline produced a rise in the oxygen consumption by adipose tissue, a rise in temperature and a release of non-esterified fatty acids into the blood. These effects on muscle and fat tissue may be coupled, in that the fatty acids liberated from adipose tissue may act as a source of fuel for the muscle tissue.

THE MODE OF ACTION OF NORADRENALINE

The mode of action of noradrenaline in increasing the metabolic rate, is not understood. It is a process of rapid activation and inactivation and it may be possible that conventional biochemical procedure cannot determine the mechanism. However, two possible mechanisms have been suggested. One possibility is that noradrenaline activates metabolic pathways of low phosphorylative efficiency, that is the pathway acts as a calorigenic shunt in which most of the energy of the substrate appears as heat. A second possibility is that there is an increased utilization of ATP without a corresponding increase in work yield. Thus if the fatty acid oxidation–synthesis cycle were to operate in such a way that there was no net synthesis or oxidation of fatty acid, then the cycle would become an ATP utilizing heat generator. This is so because a greater amount of ATP is needed to synthesize fatty acids than is generated during their oxidation.

CENTRAL MECHANISMS INVOLVED IN NON-SHIVERING THERMOGENESIS

The hypothalamus is the principal site for integration of the autonomic nervous system, concerned as it is with the regulation of body temperature, carbohydrate and fat metabolism, cardiovascular functions, etc. The hypothalamic nuclei concerned with the function of sympathetic nerves lie in the posterior and lateral parts of the region and electrical stimulation of these areas results in a massive activation of the sympathetic nervous system and the adrenal medulla. Systematic studies of the rôle of the hypothalamus in activating noradrenaline-dependent non-shivering thermogenesis have yet to be made. However, in view of the known importance of this area of the brain in regulating the sympathetic nervous system it seems probable that such a relationship will be demonstrated.

The thyroid gland and temperature regulation

The thyroid gland, lying anterior and lateral to the trachea, is the largest endocrine gland in the body. In an adult man the gland weighs about 20 g, contrasted with the pituitary gland which weighs only 0·5 g or an adrenal which weighs 5 g. The size of the thyroid gland is perhaps released to the degree and mode of storage of thyroid hormones, for so much hormone is stored that one can use the desiccated gland to treat cases of hypothyroidism. Like other endocrine glands the thyroid has a very rich blood supply, indeed weight for weight it has a greater blood flow than the kidney, which can take about a quarter of the blood flow from the heart.

The functional unit of thyroid structure is the follicle (plate 3a), a spherical structure with acinar cells on the outside, surrounding a

cavity which is usually filled with colloid. There are two thyroid hormones released by the follicle into the blood. Thyroxine forms the
bulk of the secretion, together with tri-iodothyronine in smaller
amounts. The raw materials for the production of these two hormones
are tyrosine (an amino acid) and iodine. The thyroid has a great

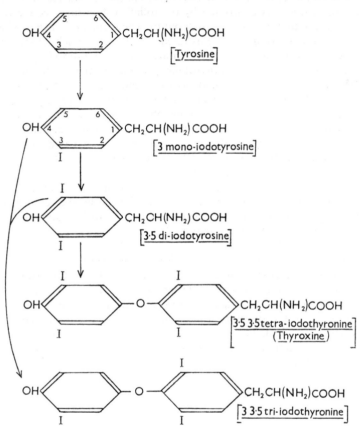

FIGURE 31. The synthesis of thyroid hormones.

affinity for iodide and it can collect and store this anion until the concentration inside the cell is at least twenty-five times that in the
circulating blood. There is an active transport mechanism, located
presumably in the cell membrane, which transports iodide from the
fluid bathing the cells into the interior of the cells. This transport
mechanism can be stimulated by the pituitary hormone TSH (thyroid

stimulating hormone), just as insulin can stimulate the active transport of glucose into the interior of cells. Inside the acinar cells of the follicle the iodide becomes loosely bound to a protein and is then secreted into the follicular cavity, where the iodide is oxidized to iodine, which then iodinates the tyrosine ring of a globulin protein. The synthesis of the hormone thus occurs in the colloid although the rate of synthesis is controlled by the follicle cells which secrete the specific globulin protein (thyroglobulin). The formation of thyroxine and tri-iodothyronine is illustrated in fig. 31. Each thyroxine molecule is made from two molecules of 3,5-di-iodotyrosine. Each molecule of tri-iodothyronine is made by the combining of one molecule of 3-mono-iodotyrosine and one of 3,5-di-iodotyrosine. Although the molecules in the diagram are represented as free compounds, it should be remembered that the tyrosine ring is actually a component of a much larger globulin molecule. Before the hormones reach the circulation the molecules of thyroxine and tri-iodothyronine are broken from the large molecules of thyroglobulin by means of a proteolytic enzyme.

The synthesis of thyroid hormones has been studied by the administration of radioactive iodide, I^{131}, followed by the removal of the gland from the animals at various intervals for the estimation of the radioactive iodinated compounds.

Factors influencing the secretion of thyroid hormones

A variety of factors influence the secretion of hormones from the thyroid gland, including the anterior pituitary, dietary components such as iodide, temperature, the adrenal glands and stress.

The rôle of the anterior pituitary in controlling thyroid function is clearly established. Following experimental removal of the anterior lobe of the pituitary, the thyroid gland atrophies and no longer secretes hormones into the blood. The administration of purified extracts of the pituitary containing a specific protein, thyroid-stimulating hormone, results in the appearance of marked changes in both the structure and function of the thyroid gland. These changes include enlargement of the thyroid with an increase in weight and vascularity. In human cases when there is overproduction of the pituitary thyrotrophic hormone the blood supply to the gland may become so great that the gland pulsates and by listening, through a stethoscope which is placed over the gland, the murmur ('bruit') of the rushing blood can be heard. In addition to these gross changes a microscopic examination of the gland shows an increase in the size of the acinar cells which surround the follicles and the appearance of a massive secretion of colloid. Biochemical changes which may be detected include, an increased uptake of iodide from the circulation and an increased rate of synthesis and release of thyroid hormones. There is evidence of a reciprocal influence of the thyroid on the anterior pituitary gland. After removal of part of the thyroid gland the anterior pituitary produces increased amounts of TSH which causes a compensatory hypertrophy of the remaining part of the thyroid and an eventual return to normal levels of thyroid hormone secretion. The rise in pituitary TSH production following partial thyroidectomy is proportional to the amount of thyroid tissue removed. This kind of reciprocal influence is called a feedback mechanism.

THE RÔLE OF THE HYPOTHALAMUS IN REGULATING THYROID ACTIVITY

If the connection of the pituitary gland with the central nervous system is adequately severed either by section of the pituitary stalk or by transplantation of the pituitary to another site in the body, then the production of TSH falls to low levels and no longer increases in response to exposing the animal to cold. It might be argued that this failure of the isolated or transplanted gland to make TSH is due to a disturbance of the vascular connections of the pituitary. However, transplanted pituitaries can still secrete large amounts of luteotrophic hormone, which indicates that the nutrition of the gland and its vascular connections are adequate.

The hypothalamico-hypophyseal portal vessels (p. 142) are the connection between the central nervous system and the activity of the anterior pituitary gland. The effects of severing these connections either by adequate stalk section or by transplantation, indicate that TSH secretion needs some neural influence. Indeed, experimental lesions in appropriate hypothalamic areas reduces the rate of TSH secretion from the pituitary, whereas electrical stimulation, particularly in the anterior regions of the hypothalamus such as the preoptic region, can induce an increased rate of release of TSH from the pituitary. There is no direct nervous link between these regions of the hypothalamus and the anterior pituitary. It is therefore presumed that a chemical factor is liberated into the hypothalamico-hypophyseal portal circulation, which stimulates the pituitary to secrete TSH. Various workers have found that extracts of the hypothalamus will induce TSH secretion from the pituitary and it is likely that the active component of the extract is a polypeptide.

Physiological effects of thyroid hormones

The thyroid hormones produce such widespread effects upon differentiation, growth and metabolism, that an absence of the hormones in man, because of a failure in thyroid development results in cretinism. A cretin is a sexually immature dwarf with the mentality of an imbecile. Although the hormone may be specifically concerned in differentiation and growth, its action on metabolism, the so-called 'calorigenic action' is the basic one in the adult mammal. This calorigenic action is represented in terms of the basal metabolic rate. This term, the basal metabolic rate, describes the minimum rate of energy expenditure per unit time which will maintain life, i.e. to maintain body temperature and provide energy for the contraction of the heart, respiratory movements, etc. In the measurement of basal metabolic rate many influences on normal metabolism, e.g. exercise, digestion, are excluded. The basal metabolic rate is therefore only a fraction of the normal metabolism and it is this fraction which is influenced by thyroxine. When measurements of the basal metabolic rate are made in man, the subject of the test is in a fasting state and at rest, physically and mentally. Under these conditions almost all of the energy changes of the body take the form of heat production and the overall exchange can be described in terms of the oxidation of energy stores to liberate heat. The basal metabolic rate is thus expressed as calories per square metre of body surface per hour. An average value for a healthy adult male is 40 cal/m²/h, which amounts to about 1,700 calories per day. Total heat production can be measured directly by knowing the temperature change of a known weight of water which circulates around the walls of the chamber containing the test individual. The technical problems of this direct calorimetry are very great, since one has to measure every avenue of heat loss from the individual—to the air in the chamber, to other objects in the chamber, and loss of heat utilized in vaporization of water, etc. Indirect measurements have been devised to measure the basal metabolic rate.

The thyroid gland has such potent effects on basal metabolic rate that deviations from the normally accepted standards in man are used to detect disorder of thyroid function or to determine the dosage of drugs used to correct thyroid disorders. A deficiency of thyroid hormone in mammals produces a fall in basal metabolism. In most species the maximal reduction in basal metabolic rate following thyroidectomy is in the order of 25–35%. Studies of the oxygen consumption of isolated tissues indicate that all body tissues participate in producing this lowering of oxygen consumption. The effect of thyroidectomy on basal metabolic rate is not immediate. There is a gradual fall of basal metabolic rate, reaching minimal values in 2–3 weeks in the rat and in 40–60 days in man. The fall in basal metabolism is, of course, associated with a decrease in the utilization of oxygen and a decreased heat production so that the body temperature may be low, particularly in a cold environment. Pulse rate and cardiac output (i.e. the amount of blood expelled from the left ventricle in one minute) are both reduced, partly because of the reduced need for oxygen by peripheral tissues but also because of the reduced metabolism of the heart itself. A variety of other defects also arise, particularly in the growing animal, such as the failure of the development of the nervous system and the skeleton and disorders of skin, teeth and hair. These other defects may be secondary to the basic action on basal metabolic rate, although there is no clear evidence for this. Some drugs are capable of mimicking the action of the thyroid on basal metabolism, but they are unable to replace the effect of the thyroid hormone on growth and development. For this and other reasons an alternative view has been put forward that thyroxine has a specific effect on growth and development independent of the effects on basal metabolism.

The administration of thyroid hormone produces a rise in the basal metabolism. This is associated with increased oxygen consumption and heat production. This calorigenic action has been demonstrated in tissues isolated from animals treated with thyroid hormones during life. The effect on metabolism is manifested in a variety of ways in man, including a rise in pulse rate and cardic output, cutaneous vasodilation and sweating in response to increased heat production, increased excitability of the nervous system, increased excitability of the gastrointestinal tract, etc. Administration of thyroxine does not produce immediate effects; following a latent period in which there is no response, there is a gradual rise over several days to maximal values, followed by a gradual fall in metabolic rate. Tri-iodothyronine has a more rapid effect on metabolism, and this is probably related to the fact that it is much more loosely bound to the specific plasma globulin (which binds and transports thyroxine) and therefore tri-iodothyronine has a ready access to tissue fluids and body cells.

Mode of action of thyroxine

Early in the history of the investigations into the action of thyroxine it was shown that tissues isolated from animals, which during life had been treated with thyroxine, showed an increased rate of oxygen consumption. However, great difficulty has been encountered in trying to reproduce these findings by the application of the hormone directly to tissues *in vitro*. A large number of individual research workers have made detailed investigations into this subject, but in spite of abundant observations on the effect of thyroid hormones on cellular and subcellular systems, there is no generally accepted view about the mode of action of the hormones.

One significant effect of thyroid hormones is an 'uncoupling' of the reactions involved in oxidative phosphorylation, when the hormones are added to *in vitro* preparations of mitochondria isolated from normal individuals. In the breakdown of molecules such as glucose and pyruvic acid during respiration, energy is released. If this energy is to be available for metabolic activities it needs to be present in the form of ATP. The reactions linking the release of the energy in the oxidative processes with the ultimate appearance of that energy in the molecule of ATP, are described as coupling processes. Uncoupling describes an impairment of these energy transfer mechanisms. When thyroid hormones cause uncoupling the process of oxidative metabolism must proceed at a faster rate if enough ATP is to be made for the need of the organism. Thus a large fraction of the energy released during oxidative metabolism appears as heat and is not available for doing work. It must be said that other substances, such as dinitrophenol, are also potent uncouplers of oxidative phosphorylation, but they cannot replace thyroid hormone in the body.

The oxidation of succinic acid to fumaric acid is a well-known reaction controlled by succinic dehydrogenase, cytochrome and electron carriers (the so-called succinoxidase system). Thyroidectomy reduces the succinoidase activity of skeletal muscle in dogs by 16%.

It has also been suggested that the effect of thyroxine on the enzymes of oxidative phosphorylation is due to an interaction with metal ions, which are essential for the activity of the respiratory enzymes. The uncoupling effect of thyroxine, in vitro, can be prevented by increasing the concentration of magnesium ions, or can be augmented by reducing the concentration of magnesium ions.

The problems of providing an explanation of the effects of thyroid hormones in the body, on the basis of the multiple effects of the hormone on subcellular systems or on isolated enzymes, has led some workers to relate the action of the hormone to effects on membranes —of whole cells, nuclei or mitochondria. Thus thyroxine was found to increase the rate of swelling of mitochondria in vitro. Magnesium ions also antagonized this effect of thyroxine on mitochondria.

THE EFFECT OF ENVIRONMENTAL TEMPERATURE ON THYROID ACTIVITY

Environmental temperature has potent effects upon the secretory activity of the thyroid gland in various species. In sheep, ox and pigs, for example, in North America the thyroid gland contains three times as much iodide in June–November as it does in December–May. Removal of the thyroid gland in rats and guinea-pigs has been found to reduce their metabolic response to cold exposure. However, some metabolic response to cold persists in the absence of the thyroid gland and even when thyroxine stores are exhausted, indicating that factors other than the thyroid hormones must be operating in the response to cold exposure. Fregly observed similar effects by the administration of a drug, thiouracil, which blocks the uptake of iodide by the thyroid gland, thus reducing the rate of synthesis of thyroxine. He found that thiouracil accelerated the cooling rate of rats exposed to cold by a factor of 55%. A quantitative relationship between environmental temperature and thyroid stimulation was shown by Dempsey and Astwood. They used rats treated with thiouracil in their drinking water to suppress thyroid gland activity. Following the depression of thyroid function the pituitary is released from the inhibitory effects of thyroxine and produces increasing amounts of TSH, which causes hypertrophy of the thyroid. These workers estimated the degree of thyroid stimulation by the pituitary, under different environmental temperatures, by noting the daily dose of thyroxine which was needed to prevent the development of thyroid hyperplasia, following thiouracil treatment. At ambient temperatures of $1°$ C $9·5$ μg thyroxine/day was needed to prevent thyroid hyperplasia whereas at $25°$ C and $35°$ C the daily doses needed were 5 μg and $1·7$ μg respectively.

The central nervous system (hypothalamus) is responsible for initiating the increased pituitary–thyroid activity which occurs on cold exposure. Knigge and Bierman found that when a pituitary gland was transplanted into the cheek pouch of a hypophysectomized hamster, this temporarily restored the release of hormone from the thyroid gland. However, exposure to cold does not produce an increase in the thyroid secretion in these 'transplant' animals, whereas it does in the normal animal where the pituitary is in a normal structural relationship with the hypothalamus. The work of Anderson, Ekman, Gale and Sundsten (1963) showed that cooling of the pre-optic region of the hypothalamus in conscious goats caused a rise in thyroxine production. This response was abolished by a lesion placed in the median eminence (p. 142) and thus appears to be activated by the pituitary.

The adrenal glands and temperature regulation

Both cortical and medullary components of the adrenal glands are involved in thermoregulatory adaptation mechanisms. The rôle of the adrenal medulla as a component of the sympathetic nervous system has already been described.

During cold exposure, the adrenal cortex releases increasing amounts of 17-hydroxycorticoids into the circulation. It is probable that all degrees of cold, including prolonged exposure, produces an increase in the amount of 17-hydroxycorticoids released. Watan and Yoshida made a prolonged study of the urinary excretion of steroids in men and women, over a 14-month period. The excretion of steroids varied with the seasons, being highest in the winter months and lower in the summer. Iampietro and his colleagues found that removal of the adrenal glands from physically restrained cold-exposed rats caused a fall in the deep body temperature. This effect of adrenalectomy could be partially prevented by the administration of adrenal cortical extracts or cortisone. When untreated adrenalectomized animals are exposed to cold they rarely survive the lowering of the colonic temperature to 22° C, whereas with intact adrenals, rats can be cooled to temperatures lower than this and still survive. Doses of adrenocortical hormones which are too small to have any effect on temperature control did nevertheless have an effect on survival.

The interaction of endocrine glands in thermoregulation

It is difficult to make a precise assessment of the rôle of a particular endocrine organ in thermoregulation because of the complex interplay between the various organs. Thus Fregly observed that the rate of cooling of restrained rats is increased by 84% by adrenalectomy, 55% by treatment with propylthiouracil (equivalent to a chemical thyroidectomy) and by 94% by a combination of both treatments. This data suggests that adrenalectomy is also associated with reduced thyroid function and indicates the difficulties in assessing the relative rôles of the adrenal and thyroid glands during cold exposure. Other experimental findings indicate an inhibitory influence of the adrenal cortex on the thyroid gland. Interaction is also known to exist between the thyroid and the adrenal medulla and between the adrenal medulla and the adrenal cortex via the action of adrenaline on hypothalamic centres (p. 111).

Summary of hormonal effects on thermoregulation

Because of the complex interplay of endocrine glands in thermoregulation it is difficult, in our present state of knowledge, to give a clearly defined description of the rôle of any particular endocrine gland. Furthermore, the rôle of the endocrine glands varies not only between different species but also during the development of an individual of one species. In general, hormonal-dependent responses to cold are more a feature of small mammals particularly in the neonatal period.

The initial responses to cold are dependent upon the nervous system, and via hypothalamic connections a variety of endocrine systems are activated. There is a rapid activation of the sympathetic nervous system and adrenal medulla, and the chemical mediators noradrenaline and adrenaline, initiate not only the cardiovascular adjustments to cold but also the increases in metabolism (and heat production) in particular organs. There is also a hypothalamic activation of the anterior pituitary, and chemical messengers in the form of TSH and ACTH activate the thyroid and adrenal cortex respectively. The degree of thyroid and adrenocortical activation depends on the severity of cold exposure. Finally, there is the developing interplay of thyroid–adrenal cortex, thyroid–adrenal medulla, adrenal medulla-adrenal cortex, the real significance of which awaits elucidation.

12

The Regulation of Endocrine Secretions: the Hypothalamus and Pituitary Gland

---

The secretory activity of endocrine glands is controlled in a variety of fashions. Perhaps the simplest mechanism is one in which the product of the action of the hormone itself regulates secretory activity. For example, the net effect of the various actions of the hormone insulin is a fall in the concentration of glucose in extracellular fluids; this effect produces a direct braking action on the secretory activity of the islets of Langerhans, causing a decline in the output of insulin (p. 28). Similarly the secretion of parathormone by the parathyroid glands is determined by the plasma concentration of the calcium ion; the net effect of the physiological actions of the hormone is to cause a rise in the concentration of calcium ions in body fluids and the parathyroids respond to this change by a reduced secretion of the hormone (p. 67). A more complex type of control is one in which the secretory activity of one gland is determined by a humoral factor produced by another gland. Thus the pituitary gland regulates secretory activity of the thyroid, adrenal cortex and gonads by means of 'trophic' hormones distributed by the blood stream to act upon the particular target organs. The humoral products of the target organs may in turn act back upon the pituitary gland to alter its secretory activity, that is a feedback or servo mechanism exists.

However, such simple regulatory systems, although they may be perfectly adequate for some homeostatic functions would prove totally inadequate for the function of adaptation of the activities of the organism to the fluctuating external environment. It is not enough that the adrenal cortex, under the influence of the pituitary trophic hormone, produces adequate amounts of hormones in the healthy, well-nourished, resting individual. It must be able to vary the output of hormones to adapt the organism to fasting, disease and stresses of all kinds. Similarly it is not enough that the ovaries are maintained in a healthy state, producing sex hormones and ova, but in those animals having seasonal breeding in temperate latitudes these functions need to be harnessed to changes in the external environment, so that the young are produced under favourable conditions of climate and food supply.

Vertebrates have evolved two major systems subserving the functions of integration and adaptation—the nervous and endocrine systems. The development of links between these two organizations enables the activities of endocrine tissues to be functionally related not only to fluxes in the internal environment but also to the external world with its changing conditions of light, temperature, sound, smell, contact, etc. Perhaps the simplest relationship between nervous and endocrine tissue is that of the sympathetic nervous system and the adrenal medulla. The activity of the medulla is virtually completely controlled by the sympathetic nervous system and when the adrenal glands are denervated, then the secretion of hormones falls to a low level which does not increase under those circumstances in which there is normally an outpouring of adrenal medullary hormones into the circulating blood (p. 112). The normal adrenal medulla is activated whenever there is a widespread discharge of activity in the sympathetic nervous system and the medullary hormones reinforce and elaborate the action of those same chemical substances liberated in much smaller and discrete amounts at the terminals of sympathetic postganglionic neurones. The two systems, nervous and endocrine, are synergistic. In fact, one can regard the adrenal medullary cells, and also the widely distributed extra-medullary chromaffine tissue (p. 102), as highly modified components of the sympathetic nervous system itself which have ceased to be concerned in the transmission of nerve impulses and have become specialized in producing in large quantities the chemical transmitter substances of the sympathetic nervous system.

The most significant relationship between endocrine and nervous tissue lies in the floor of the fore-brain and it is this connection which will be the main subject of this chapter. The anterior pituitary gland regulating as it does the activities of the thyroid gland, ovaries, testes, adrenal cortex and mammary glands, has been rightly named the 'master gland' or 'the conductor of the endocrine orchestra'. Removal of the anterior lobe of the pituitary results in a failure of growth and reproduction and the onset of metabolic disturbances caused by deficiencies of thyroid and adrenocortical hormones, which ultimately prove fatal. To some extent the activities of the anterior pituitary are regulated by the secretory products of the organs which it controls; thus high blood levels of thyroxine depress the secretion of pituitary thyroid-stimulating hormone, and this ultimately leads to a decline in production of thyroxine by the thyroid gland, i.e. a negative feedback mechanism exists. However, it has already been stressed that this type of control is not adequate to adapt the animal to external circumstances. Some link of the pituitary gland with the external environment via the nervous system has long been postulated. Only such a link can explain the influence of an enormous variety of external factors on the secretory activity of endocrine glands controlled by the pituitary. These factors include, for example, the effects of day length, temperature, odours and mating stimuli on sexual rhythm and functions, and the effect of a vast range of 'stressful' situations on adrenocortical function (p. 119), or the responses of the thyroid gland to changes in external temperature (p. 136). Perhaps the clearest example one could cite in support of the view that CNS connections of the pituitary gland are a prerequisite of adaptive behaviour is the case of those female mammals which release ova from the ovaries only after the act of mating. The nervous stimuli associated with mating influence the anterior pituitary gland causing it to release luteinizing hormone into the circulating blood which initiates ovulation. This effect of mating is abolished if the connection of the pituitary gland with the central nervous system—the pituitary stalk—has previously been severed.

Pituitary CNS connections (fig. 32)

It seems clear that the anterior lobe of the pituitary must receive its influences from the nervous system by way of the pituitary stalk. But it has been known for many years that the anterior pituitary has, in fact, no nervous connection with the brain. There is, however, an unusual arrangement of blood vessels supplying the pituitary—the hypothalamico–hypophyseal portal system—first described by Popa and Fielding in 1930. A portal system of blood vessels is one in which there are two distinct capillary beds interposed between artery and vein. In this portal system the first capillary bed lies at the upper end of the pituitary stalk in the part of the hypothalamus called the median eminence, and it is connected to the second capillary bed in the anterior lobe itself. Direct observations have shown that the blood circulation is from the median eminence, down the stalk, to the anterior lobe. Thus the pituitary has a special vascular connection with the floor of the fore-brain, the hypothalamus. The area of contact between the pituitary and hypothalamus is not limited to the

median eminence; the axons of neurones located widely in the hypothalamus converge on the pituitary stalk to terminate on the surface
of the median eminence.

HYPOTHALAMIC-RELEASING FACTORS

It seems certain that the hypothalamus exerts its influence on the
anterior pituitary gland by means of chemical substances liberated at

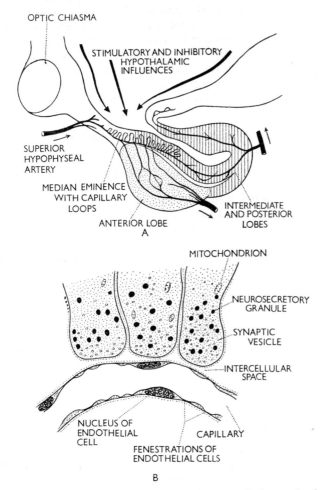

FIGURE 32. (A) Vertical section through the pituitary gland and
hypothalamus showing the hypothalamico–hypophyseal portal circulation. (B) Diagram made from electron micrographs of the median
eminence showing three nerve terminals on a blood capillary.

nerve terminals on the median eminence. These substances enter the blood in the portal system to be transported to the anterior lobe where they exert effects on the secretion of pituitary trophic hormones. The area of contact between nerves and blood vessels in the median eminence is increased by the presence of abundant capillary loops. Electron micrographs of this region show that nerve terminals contain masses of neurosecretory material and synaptic vesicles. It is presumed that the chemical regulators of the anterior pituitary are liberated from the nerve terminals into the extracellular space before gaining access to the blood in the capillaries of the portal system. Indeed, transmission of substances is facilitated by the character of the capillary wall which is 'fenestrated', that is the wall has gaps in the endothelial lining of the vessel, a feature typical of capillaries in regions of active absorption or secretion (fig. 32B).

Experimental evidence for hypothalamic influences on anterior pituitary function

The evidence of a vascular connection of special character between the median eminence and the anterior lobe provides the morphological basis for hypothalamic-pituitary influence. The experimental verification that this indeed provides a functional connection can be considered under various headings.

The effects of surgical stalk section or transplantation of the anterior pituitary gland

Early work to determine the significance of the connection of the anterior lobe to the central nervous system by way of the pituitary stalk consisted mainly in the severance of this connection either by surgical section of the pituitary stalk or the removal of the pituitary gland and its transplantation to some other site in the body. These operations must be carefully performed to obtain valid results. The operation of stalk section produces only temporary separation of the hypothalamus and pituitary unless precautions are taken to prevent the regrowth of capillary blood vessels from the median eminence into the stalk. The insertion of a barrier of some inert material, e.g. a disc of mica, between the median eminence and the stalk prevents the re-establishment of vascular connections. If such a precaution is not taken the regeneration of portal blood vessels may occur within 24 hours of stalk section. If the operation of stalk section or transplantation is adequately performed, then it is followed by a variety of defects of bodily functions. Atrophy of most of the adrenal cortex occurs with the associated failure of adequate responses to stress. The ovaries and testes atrophy and sexual activity disappears. These and other changes occur even although the experimentally isolated pituitaries are apparently healthy—judged from their histological appearance—and are well vascularized. The defects in function must be attributed to a loss of connection with the hypothalamus and this implies that the hypothalamus must normally exert a stimulatory effect of anterior pituitary function.

One remarkable feature of the pituitary gland separated from connection with the hypothalamus is the production of large amounts of the hormone prolactin (luteotrophic hormone) which contrasts with the reduced secretion of TSH, ACTH, and gonadotrophic hormones. In fact, the isolated pituitary secretes more prolactin than the normal pituitary gland. These findings indicate that in addition to a stimulatory influence of the hypothalamus on the pituitary there is an inhibitory influence and it must be assumed that in the normal animal prolactin production is inhibited by hypothalamic influence until this influence is curbed by other factors.

The effects of brain extracts and various known biologocal activators on anterior pituitary function

The evidence derived from surgical isolation of the pituitary gland supports the view that the hypothalamus exerts potent influences on anterior lobe function and that these may be of stimulatory or inhibitory nature. Because of the absence of any nervous link with the hypothalamus it seems likely that the pituitary is regulated by means of chemical activators liberated from nerve terminals in the median eminence and transported to the pituitary in the hypothalamico-hypophyseal portal system.

Extracts made from the median eminence are particularly potent in causing release of hormones from the anterior pituitary gland. Some active components of these extracts have been isolated and purified. The substances are of a polypeptide nature and they exert specific effects. One polypeptide stimulates the release of ACTH (CRF—corticotrophin releasing factor) and another causes release of TSH (TRF—thyrotrophin releasing factor). Campbell and co-workers (1961) also described a stimulation of secretion of luteotrophic hormone from the pituitary gland following the injection of extracts of median eminence into the pituitary gland of rabbits. These workers inserted a needle through the skull directly into the pituitary and injected the extract into the gland (fig. 34). Injection into the pituitary of an extract of as little as 2·5 mg of fresh tissue caused ovulation (the result of LH secretion from the pituitary) in eleven out of twenty-two rabbits, whereas the injection of an extract of 100 mg of tissue into a peripheral vein elicited ovulation in only one out of seven animals. Extracts made from other areas of the brain (cerebral cortex, caudate nucleus and corpus callosum) were ineffective when injected into the pituitary and so were a whole range of biologically active substances, including histamine, 5-hydroxytryptamine, oxytocin, vasopressin, substance P and adrenaline. Using the same technique of pituitary injection purified CRF was found to have no effect on ovulation. This indicates that the luteinizing hormone-releasing factor (LRF) is distinct from the polypeptide which stimulates the release of ACTH from the pituitary.

Recently it has been possible to collect blood from the hypothalamico-hypophyseal portal system after approaching the pituitary gland

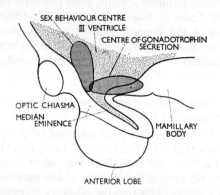

FIGURE 33. Vertical section through the pituitary gland and hypothalamus in the midline showing the gonadotrophic and sexual behaviour centres in the cat.

by way of the posterior pharyngeal wall (Worthington, 1966). This blood has been shown to contain substances which do not occur, at least in similar concentrations, in blood obtained elsewhere in the body. It has been shown that when injected into rats this blood is capable of stimulating the release of ACTH and luteinizing hormone from the anterior pituitary gland.

Electrical stimulation and ablation experiments

If the hypothalamus regulates anterior pituitary function one would anticipate that electrical stimulation or destruction of areas in the hypothalamus would modify anterior lobe function. And indeed electrical stimulation of various regions of the hypothalamus has been found to produce an increase in the secretion of various pituitary hormones. Thus stimulation of the posterior region of the hypothalamus causes a discharge of ACTH and LH from the anterior lobe. In the anterior region of the hypothalamus stimulation results in an increased secretion of thyrotrophic hormone. Ovulation also results from localized hypothalamic stimulation.

Localized lesions of the hypothalamus produced, e.g. by electrocautery, also disturbs anterior lobe function. Hypothalamic lesions have been found to block ovulation in various mammalian species. The areas regulating ovulation are located in different regions of the hypothalamus in different species. Some hypothalamic lesions can also increase the activity of the anterior pituitary; thus lesions placed in the anterior hypothalamus can precipitate precocious sexual development in immature female rats or produce sexual activation in

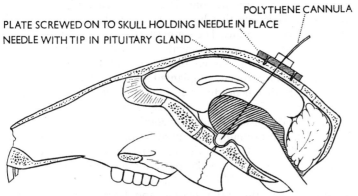

POLYTHENE CANNULA
PLATE SCREWED ON TO SKULL HOLDING NEEDLE IN PLACE
NEEDLE WITH TIP IN PITUITARY GLAND

FIGURE 34. V.S. rabbit skull showing the technique of intra-pituitary injection.

ferrets during the winter months when the animals are normally sexually inactive.

These experiments indicate that hypothalamic 'centres' exert both inhibitory and excitatory influences on the anterior pituitary. It is possible to map out rough areas of hypothalamic function, particular areas being concerned with particular functions (fig. 33). However, this task is beset with the problems of species differences in the geography of hypothalamic functions and the experimental problem of destroying completely a particular area of the hypothalamus without damaging adjacent cells or fibre tracts. Some of these fibre tracts arise not only in the hypothalamus itself but also from other parts of the nervous system, including the brain stem reticular formation, the amygdala, hippocampus and orbito-frontal region of the cerebral cortex. One must regard the pituitary stalk as the final common path of many nervous influences both hypothalamic and extra-hypothalamic.

The influence of hormones on the nervous system

The relationship between the hypothalamus-pituitary gland and the peripheral target organs is a reciprocal one. It seems fairly clear that the pituitary gland and its target organs are coupled together by feedback loops of negative character. Thus if there is a rise in the concentration of hormones produced by the target organs (adrenal cortex, thyroid, gonads) these secretions exert a direct braking action on the secretion of trophic hormones by the anterior pituitary. Experimental observations confirm that the anterior pituitary is directly sensitive to thyroxine and adrenocortical hormones, but is probably influenced by sex hormones only in excessive quantities.

It also seems certain that hormones, particularly sex steroids, influence the hypothalamus. The simplest explanation of this feedback mechanism is that there are elements in the hypothalamus which are directly sensitive to hormones in the circulating blood. Thus if a small amount of oestrogen is implanted into the anterior region of the hypothalamus, then it produces an inhibitory effect on the pituitary secretion of gonadotrophins, an effect 125 times greater than if the hormone is administered subcutaneously. The effects of implants of oestrogen is profoundly influenced by the site of implantation in the hypothalamus. Thus Davidson and Sawyer implanted minute amounts of oestradiol into the posterior median eminence—basal tuberal region of the rabbit hypothalamus and found this produced a failure of ovulation after mating and later an atrophy of the ovaries. When these amounts of oestradiol were implanted into the anterior pituitary gland or in other regions of the brain, then no effects on ovulation were observed. This type of experiment indicates that there are particular areas of the hypothalamus which are very sensitive to sex steroids.

Histological studies have shown that the administration of sex steroids produces observable morphological changes in the nerve cells of the hypothalamus. In the rat either an increase or decrease in the blood level of sex steroids results in a decrease in nuclear size of nerve cells lying in the anterior hypothalamus. Using radioactive oestrogen (H[3]–hexoestrol) administered subcutaneously Michael (1962) was able to localize, by autoradiography (p. 15), areas of nerve cells in the hypothalamus which selectively accumulate oestrogen. When radioactive oestrogen is administered to an animal a variety of 'non-target tissues' such as muscle or thyroid gland take up the hormone, reaching a low maximal uptake after about 2 hours after which radioactivity is lost. However, other 'target tissues' such as uterus or vagina continue to accumulate radioactive oestrogen for 5–8 hours and retain this oestrogen for 6–12 hours or more. Into this class of target tissues falls the hypothalamus and pre-optic areas of the brain. The remaining areas of the brain fall into the class of 'non-target tissues'.

Hypothalamic feedback mechanisms and sexual behaviour

The steroid sex hormones produced by the ovary and testis not only influence those hypothalamic mechanisms which modulate anterior pituitary function but they also affect those areas of the hypothalamus concerned with sexual behaviour. Specific areas of the hypothalamus concerned in sexual behaviour have been localized by the production of localized hypothalamic lesions.

The administration of oestrogen to a female cat induces the appearance of mating behaviour. This effect of oestrogen is seen even in the absence of the pituitary gland, but is lost after the production of lesions in the anterior hypothalamus. The position of hypothalamic areas determining sexual behaviour vary from species to species. Further, these areas are not identical with those which regulate the secretion of gonadotrophins from the anterior pituitary. If the effect of sex steroids on sexual behaviour is due to a direct effect of the hormones on specific nerve centres it should be possible to induce the appearance of sexual behaviour by the implantation of hormones into appropriate areas. The experimental use of implants of solid hormone pellets or crystals has proved to be of considerable use in the localization of these areas since unlike aqueous or oily solutions these solid deposits produce mainly local effects. In the rat, for example, Lisk implanted solid oestradiol by means of fine steel tubing inserted into the brain. He was able to localize two areas in the hypothalamus sensitive to the effect of this oestrogen. At one focus implants of oestrogen inhibited the secretion of gonadotrophins by the anterior pituitary and at another focus oestrogen brought the animal into oestrus behaviour.

The sexual differentiation of the hypothalamus

The pituitary gland of adult male and female animals differs considerably in the pattern of gonadotrophin secretion. In female animals there is a rhythmical secretion of gonadotrophins so that ovaries undergo the cyclic processes of follicular development under the influence of FSH, followed by ovulation and corpus luteum formation under the influence of LH. In the male, however, there is a relatively steady production of gonadotrophins during the breeding season. The causes of these differences in hormone secretion does not lie in the pituitary gland itself. If the pituitary gland of an adult female is removed and replaced by the pituitary gland of a male animal, then the transplanted pituitary changes its pattern of gonado-trophin secretion into that of the female type and the oestrous cycle is maintained. The sex difference is located in the hypothalamus and it is the hypothalamus which determines the pattern of gonadotrophin secretion from the pituitary gland.

In the rat and other mammals the hypothalamus of the newly born animal is apparently sexually indeterminate and is capable of differ-entiating into either male or female type irrespective of the sex of the

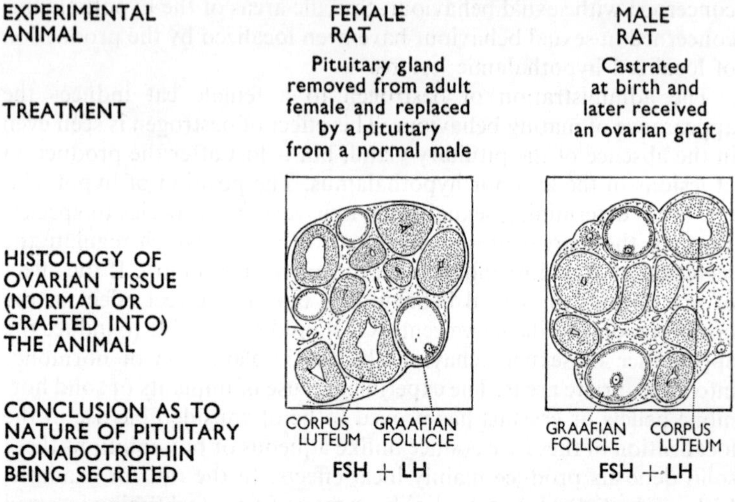

EXPERIMENTAL ANIMAL	FEMALE RAT	MALE RAT
TREATMENT	Pituitary gland removed from adult female and replaced by a pituitary from a normal male	Castrated at birth and later received an ovarian graft
HISTOLOGY OF OVARIAN TISSUE (NORMAL OR GRAFTED INTO) THE ANIMAL		
CONCLUSION AS TO NATURE OF PITUITARY GONADOTROPHIN BEING SECRETED	CORPUS LUTEUM GRAAFIAN FOLLICLE FSH + LH	GRAAFIAN FOLLICLE CORPUS LUTEUM FSH + LH

FIGURE 35. The use of ovaries (normal or transplanted) as indices of the ments in the rat.

gonads, provided that it is given the appropriate conditions. If the ovaries or testes are removed from newly born animals the hypo-thalamus continues to differentiate into the female type capable of initiating sexual rhythms and behaviour of female type provided that ovarian tissue is grafted into the animal at a later date. The main factor which determines the direction of differentiation of hypo-thalamic function is the presence or absence of male sex hormone. If male sex hormone is administered early in post-natal life, then the hypothalamus differentiates into the male type irrespective of the genetic or gonadal sex of the individual. Thus if a newly born female rat is treated with androgen, the hypothalamus differentiates into the male type. When sexual maturation eventually occurs, the ovary shows marked follicular development but ovulation and corpus euteum formation does not occur. If a male rat is castrated at birth or early in post-natal life, then the hypothalamus differentiates into the female type in the absence of male hormone (fig. 35).

The hypothalamus and the onset of sexual maturation

Not only does the hypothalamus regulate anterior pituitary function and determine sexual behaviour but it is also concerned in regulating the timing of sexual maturity. Normally there is a latent period between the differentiation of the gonads and reproductive tract and the assumption of full sexual activity, and the length of this latent period differs considerably from one species to another. In young female rats

FEMALE
RAT
Treated with
androgen
at birth

MALE
RAT
Castrated as
adult and received
an ovarian graft

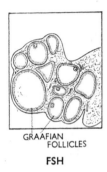

GRAAFIAN
FOLLICLES

GRAAFIAN
FOLLICLES

FSH

FSH

pattern of the secretion of pituitary gonadotrophins after various treat-

lesions placed in the anterior hypothalamus accelerate the onset of sexual maturation. It is assumed that the hypothalamus of the young animal is very sensitive to sex hormones and responds to the low levels of circulating steroids of the sexually immature animal by exerting a restraining action on the secretion of gonadotrophins by the anterior pituitary gland. The experimental effect of lesions placed in the anterior hypothalamus is postulated as an effect of removal of sex hormone-sensitive areas in this region which exert inhibitory influences on the secretion of gonadotrophic hormones by the anterior pituitary gland. It is presumed that during normal sexual maturation some as yet unknown maturation process in the anterior hypothalamus lessens its sensitivity to sex hormones. This maturation process removes the inhibitory influences from the anterior pituitary which responds by an outpouring of gonadotrophic hormones and the maturation of the gonads.

13

Hormones and the Integration of Gastrointestinal Activity

The integration of motor and secretory activity in the gut

The need for co-ordination and control

Digestion depends upon the secretion of many enzymes into the alimentary canal so that food may be hydrolysed and prepared for absorption in the ileum. The synthesis and secretion of these proteinaceous enzymes involves the expenditure of energy and the use of valuable raw materials. It is vital to the economy of the body that these enzymes are secreted only when appropriate food materials are present in the alimentary canal. Digestive enzymes must be secreted in an appropriate sequence and act for the right length of time, in the most suitable concentration in a medium of suitable pH, if economical and efficient digestion is to be achieved. Food in the lumen of the gut must be propelled from one section of the gut to the next at a suitable speed and in a co-ordinated manner. This co-ordination of secretion and movement is brought about by the actions of nerves and hormones.

Salivation

Food in the mouth stimulates the sense organs, e.g. taste buds on the tongue, which discharge impulses via nerves to the salivary centre in the medulla of the hind brain. Secreto-motor impulses are discharged from the medulla via parasympathetic nerves and cause saliva to flow from the salivary glands. Thus salivation is under nervous control.

Pavlov, in the late 19th century, showed that objects in the environment can act at a distance on the salivary glands. The sight of food will cause a hungry dog to salivate long before the food comes into contact with the taste buds. Even the sight of the person who usually

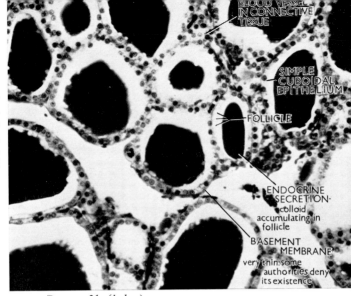

PLATE 3a (*right*)
)tomicrograph. T.S. thy-
d gland of monkey.
ag. 380 ×). From Free-
n, W. H. and Bracegirdle,
B., *An Atlas of Histology*.
inemann Educational
oks Ltd., 1966.

PLATE 3b (*below*)
T.S. testis of cat showing seminiferous tubules and interstitial tissue. (Mag.
400 ×). From Freeman, W. H. and Bracegirdle, B. B., *An Atlas of
Histology*. Heinemann Educational Books Ltd., 1966.

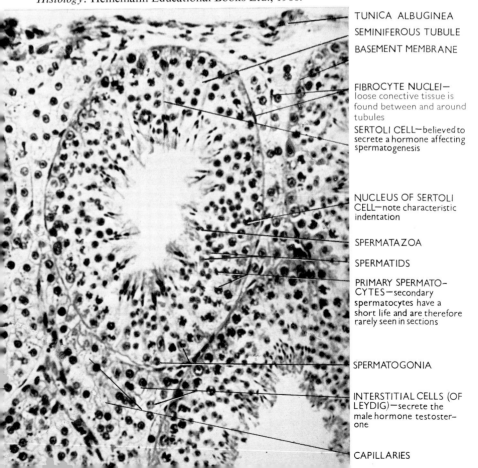

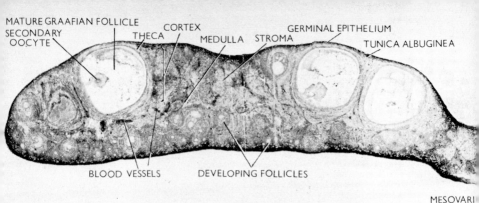

MATURE GRAAFIAN FOLLICLE
SECONDARY
OOCYTE
THECA
CORTEX
MEDULLA STROMA
GERMINAL EPITHELIUM
TUNICA ALBUGINEA

BLOOD VESSELS   DEVELOPING FOLLICLES

MESOVARI

PLATE 4

The structure of the ovary.
a. L.S. ovary of rabbit showing Graafian follicles. (Mag. 12 ×).
b. Section of human ovary showing the corpus luteum of pregnancy. (Mag.
3 ×). From Freeman, W. H. and Bracegirdle, B. B., *An Atlas of Histology.*
Heinemann Educational Books Ltd., 1966.

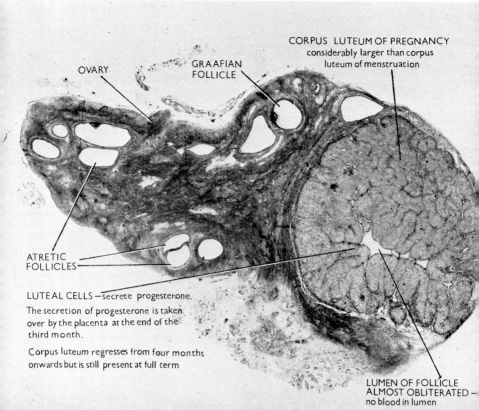

OVARY

GRAAFIAN
FOLLICLE

CORPUS LUTEUM OF PREGNANCY
considerably larger than corpus
luteum of menstruation

ATRETIC
FOLLICLES

LUTEAL CELLS —secrete progesterone.
The secretion of progesterone is taken
over by the placenta at the end of the
third month.

Corpus luteum regresses from four months
onwards but is still present at full term

LUMEN OF FOLLICLE
ALMOST OBLITERATED —
no blood in lumen

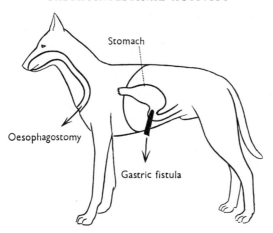

FIGURE 36. Dog prepared for sham feeding and sampling of stomach juice.

feeds the dog will initiate salivation. The dog learns to associate the sight of the keeper with the subsequent arrival of the food. The visual stimulus becomes associated with the taste of the food and is said to be an associated stimulus. The response of the dog to stimuli associated with food, rather than to the food itself, is said to be a conditioned response. These conditioned reflex responses involve the cerebral cortex and allow intelligent animals to anticipate the arrival of food by salivating before the food is eaten.

Secretions of the stomach

I. The cephalic phase

In a classical experiment first reported in 1895, Pavlov and Mme Schumov-Simanovskaja provided evidence that the vagus nerve is a secretory nerve for the gastric glands, and that a vagal reflex is responsible for controlling the secretion of gastric juice at the beginning of a meal. They prepared dogs with an oesophagostomy so that the food swallowed never reached the stomach. The dogs were provided with a gastric fistula so that the gastric juice could be collected (fig. 36). On sham feeding the dog there was a latent period of about 5 minutes, followed by a rapid flow of stomach juice which continued for as long as 3–4 hours, the amount of juice flowing gradually diminishing as time went on. This response to sham feeding can be abolished by section of the vagus nerve, or by injecting atropine intravenously. The vagus nerve produces its effects by the release of

F

the transmitter substance acetylcholine at the nerve terminals in the stomach. In this, and some other situations, e.g. smooth muscle and heart, atropine combines with the tissue 'receptors' for acetylcholine and prevents the normal activation of the tissue by acetylcholine. The juice obtained from the stomach after vagal excitation is copious in amount, and high in acid content and peptic power. In man a similar mechanism exists and the juice so formed is referred to as the appetite juice. It flows for about half an hour at the beginning of a meal.

II. The gastric phase

As the appetite juice acts on the food in the stomach, the acid and pepsin begin to hydrolyse the protein forming peptones, which accumulate in the pyloric antrum (fig. 36). In response to the presence of peptones, the wall of the antrum releases a hormone, gastrin (the antral hormone), which enters the blood stream and stimulates the glands of the fundic stomach to secrete more gastric juice. The acid juice so formed contains no pepsin. Excess acid in the pyloric region inhibits the secretion of the hormone gastrin and thereby prevents the release of acid juice from the glands of the fundus. Here, then, is a self-regulating mechanism ensuring the presence of just the right amount of acid in the stomach for the digestion of protein by pepsin.

Distension of the pylorus stimulates the flow of gastrin and a consequent secretion of acid-rich juice from the fundus. Thus more gastric juice flows just after a meal when there is much digestion to be done than 2 hours later when the pyloric antrum is almost empty.

Local vagus nervous reflexes may be involved in controlling the secretion of the hormone gastrin. Certainly an intact vagus allows the gastrin secreted to produce a greater volume of acid juice. Both the vagus nerve and the hormone gastrin work together in the intact animal to control the secretion of acid juice.

III. The intestinal phase

The duodenum monitors the composition of the food leaving the pylorus and may stimulate the flow of more gastric juice or may halt the flow.

(a) Stimulation of secretion of more gastric juice

The presence of certain foodstuffs in the duodenum stimulates the secretion of gastrin by the duodenum in the region of Brunner's glands. However, the hormone is thought not to arise in these glands. The secretion of Brunner's glands is mucoid and serves to protect the duodenal epthelium from the highly acid stomach juice. The hormone gastrin which is secreted by the duodenum into the circulating blood provokes the flow of acid-rich juice from the fundic glands of the stomach. It is considered that innervation by the vagus nerve is necessary for the full response of the fundus to gastrin and perhaps also for the release of gastrin from the duodenum.

(b) *Inhibition of the flow of gastric juice*

1. The pH of the duodenal contents is maintained at about pH 4. If too much acid is leaving the stomach, then the pH of the duodenal contents falls below this value, and when the duodenal pH falls below 2·5, then the flow of acid from the glands of the fundus ceases. This response is considered to be mediated by a reflex involving the vagus nerve.

2. The duodenum is sensitive to the osmolarity of the duodenal contents. Hypertonic solutions cause an inhibition of the flow of gastric juice.

3. Fats in the duodenum are highly effective in releasing a hormone, enterogastrone from the duodenum which inhibits the secretion of gastric juice by the stomach. An old remedy for the treatment of peptic ulcers was that the patient was recommended to drink olive oil. When the olive oil passed into the duodenum it would suppress the activity of the acid-producing cells of the stomach.

SUMMARY OF THE CONTROL OF STOMACH SECRETIONS

Many physiological mechanisms are involved in the control of the output of pepsin, water, mucus and acid into the stomach. These mechanisms include nervous reflexes mediated by the vagus nerve, the hormones gastrin and enterogastrone, and special receptors sensitive to the pH and osmolarity of the duodenal contents. The co-ordinated activity of these special mechanisms create conditions in which the digestion of protein can proceed effectively in the stomach. They also ensure that the chyme leaves the stomach at a rate which does not disturb the functions of the duodenum.

The control of movements of the stomach

Studies utilizing X-rays and electronic pressure transducers have shown that the pyloric sphincter remains open at all times. The rhythm and intensity of the peristaltic waves of contraction in the stomach vary and not all of the waves reach the pylorus. When the waves are strong and reach the pylorus the liquid part of the stomach contents is squeezed into the duodenum. The rhythm and intensity of the peristaltic waves in the stomach are controlled by the duodenum in two ways:

1. By the enterogastric reflex, a nervous reflex mediated by the vagus nerve. Distension of the duodenum initiates this reflex which results in the slowing of the rate of stomach emptying.

2. By a humoral reflex involving the hormone enterogastrone. This hormone is secreted by the duodenum in response to the presence of fats in the lumen, and the hormone produces an inhibitory effect on stomach emptying. This effect of fats in delaying the emptying the stomach accounts for the fact that meals with a high fat content satisfy hunger for a longer period than meals which have a low fat content. The sensation of hunger is partly the result of an empty stomach and anything which delays stomach emptying will avert the sensation of hunger.

Many factors which result in slowing the emptying of the stomach have been recorded, but their mode of action is not always clear. These factors include hypertonic or hypotonic solutions in the duodenum, the presence of large or irritating particles in the duodenum, and various foodstuffs, e.g. fat or excess peptone. Excess acid leaving the stomach also slows the rate of stomach emptying.

The acid contents of the duodenum are neutralized by the alkaline secretion of the pancreas. If the stomach emptied too quickly, then a situation could arise when the pancreatic juice would be unable to cope with the excess acid and the pancreatic enzymes which operate best in an alkaline medium would operate in sub-optimal conditions. By having mechanisms which slow the rate of stomach emptying triggered off by such stimuli as excess acid in the duodenum, the conditions in the duodenum are maintained within limits which are controllable.

Thus the pylorus, the sphincter and the duodenum act as a physiologically integrated unit passing food at the appropriate rate and in the proper physical and chemical conditions into the small intestine, where it meets pancreatic and intestinal secretions.

The effect of distension of the ileum on stomach emptying is important in ensuring that food is passed into the small intestine at a suitable rate. When the stomach is distended during the course of a meal the ileum increases its motility. Here is another example of an advanced warning system in operation, whereby the parts of the gut are prepared for the food to arrive. When chyme reaches the ileum there is a reflex inhibition of gastric motility. Thus the stomach is prevented from disgorging its contents into a small intestine already coping with a quantity of chyme. The splanchnic nerves are involved in mediating these reciprocal actions of the stomach and ileum.

The control of the flow of bile by cholecystokinin

The presence of fat in the lumen of the duodenum induces the duodenum to release a hormone into the blood stream. This hormone is called cholecystokinin and its target organ the gall bladder which responds by contraction and the expulsion of its contained bile into the intestine. Thus bile salts which are essential for the efficient digestion of fat are released into the intestine only when fat is present. If fat appears in the duodenum faster than the bile and pancreatic lipase can deal with it, then the rate of stomach emptying is slowed in response to the release of the hormone enterogastrone from the duodenum.

Secretion of pancreatic juice

CONTROL BY THE VAGUS NERVE

In a fasting dog the stomach contains no food and the acid stomach juice is discharged infrequently into the duodenum. When such a dog is given a meal the pancreas slowly begins to secrete pancreatic juice which is rich in lipases, proteases and amylase. This slow flow of viscid juice is attributed to the action of the vagus nerve on the pancreas. As early as 1888 Pavlov and his pupils stimulated the vagus nerve in the neck of a dog which had a pancreatic fistula, and they observed a free flow of juice from the fistula after vagal stimulation. It is realized now that this procedure would also stimulate the secretion of gastric juice and as this juice emptied into the duodenum it would initiate the humoral reflex which is described below. It is the humoral reflex which is responsible for the large flow of pancreatic juice. However, Pavlov followed his early experiments with others in 1896 in which he blocked the entrance of stomach juice into the duodenum prior to stimulation of the vagus nerve. The slow flow of enzyme-rich viscid pancreatic juice which followed demonstrated that the vagus nerve was a secretory nerve to the pancreas. The vagus influences the secretion of enzymes into pancreatic juice rather than the secretion of water and salts.

CONTROL BY THE HORMONES SECRETIN AND PANCREOZYMIN AND GASTRIA

As soon as acid and the products of protein digestion enter the duodenum, a hormone, secretin, is released from the mucosa of the duodenum into the blood stream. The pancreas responds by producing a copious flow of water and salts (e.g. sodium bicarbonate). The juice produced in response to secretin is low in enzyme content (fig. 37A). This alkaline secretion serves to neutralize the acid chyme in the duodenum. Secretin is released from the duodenum when free acid is present and the pH is four or lower. The fact that the presence of acid in the duodenum causes secretion of a watery juice from the pancreas was known as early as 1825, but it was not until 1902 that Bayliss and Starling attributed this to the effect of a hormone secreted by the duodenum. Secretin was the first hormone to be discovered in relation to the alimentary tract. Pure preparations of secretin are now available and their use has shown that the hormone also causes the liver to secrete water and bicarbonate into the hepatic bile.

During the later stages of a meal an enzyme-rich juice flows from the pancreas. The release of these enzymes is considered to be a response to the secretion of another hormone, pancreozymin, from the duodenal mucosa (fig. 37B). This hormone was discovered in 1943 by Harper and Raper in England. Pancreozymin is concerned with the transfer of enzymes across the cell membranes in the pancreas, but not with the intracellular synthesis of the enzymes.

We also know that gastrin is a strong stimulant of pancreatic enzyme output, and a weak stimulant of pancreatic flow and of bicarbonate output.

The control of intestinal secretion

In 1904 Delezenne and Frouin observed that extracts of the ileum when injected intravenously caused a stimulation of secretion of

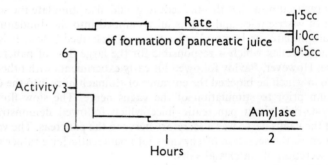

FIGURE 37. Factors affecting the secretion of enzymes from the anaesthetized cat. In the three figures the rate of secretion of pancreatic juice and its amylase activity are plotted against time. A The effect of secretin only; there is a steady output of juice with low amylase content. B The effect of the addition of pancreozymin to secretin; there is no change in the volume of secretion but there is a marked change in its amylase activity. C The effect of stimulation of the vagus nerve during the administration of secretin; there is minimal change in the rate of secretion of pancreatic juice but there is a marked increase in its amylase activity. (The gradual decline in amylase activity during the period of vagal stimulation is probably due to 'fatigue' of the preparation.) Figure reproduced by courtesy of Professor A. A. Harper from Harper, A. A. & Mackay, I. F. S. (1948). The effects of pancreozymin and of vagal nerve stimulation upon the histological appearance of the pancreas. *J. Physiol.*, 107, 89–96.

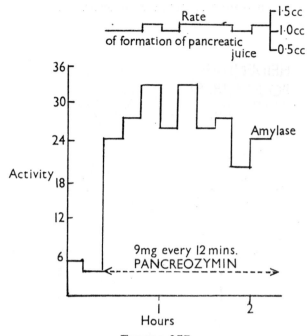

FIGURE 37B

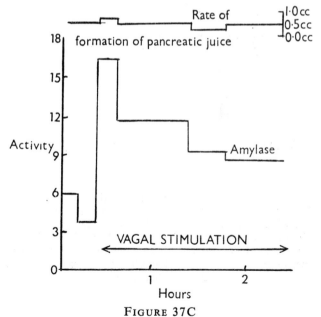

FIGURE 37C

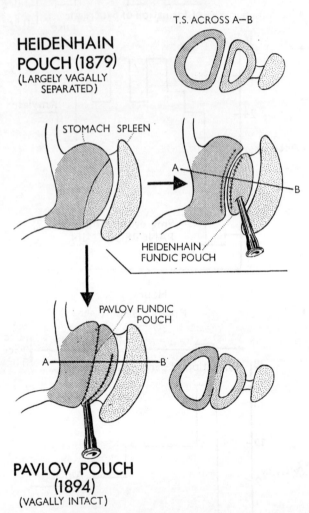

FIGURE 38. Diagram showing the preparation of the Heidenhain pouch and the Pavlov pouch of the fundic stomach. The Pavlov pouch retains an intact vagus nerve supply but in the Heidenhain pouch the main branches of the vagus nerve are cut. However the Heidenhain pouch may have had some vagal innervation, not by way of the main branches of the vagus from the stomach but via smaller nerves direct from the coeliac ganglion.

succus entericus. It has been postulated that these extracts contain a humoral substance. The active principle, called enterocrinin, has not yet been isolated in anything like a pure form and further chemical investigation of extracts having enterocrinin activity is needed.

A look at the evidence concerned in the gastrin controversy

So far this chapter has merely laid out the major items of what we believe to be true of the integrative influences, especially the hormones, of the gut. We can now look a little more closely at the historical development of our ideas of how a few of these mechanisms operate.

Pavlov suggests a nervous reflex mechanism

Pavlov was aware that the secretion of stomach juice was stimulated by the presence of food in the mouth, but he also realized that other factors were involved in the control of gastric secretion. He devised experiments, the results of which led him to believe that the pylorus was sensitive to contact with meat extract, and that when meat entered this part of the stomach, gastric juice was secreted. In these experiments he devised a surgical technique to produce a pouch of the main body of the stomach, which opened to the outside of the body. This pouch of the stomach opening to the exterior has been called the Pavlov pouch and was carefully prepared with the vagus nerve intact (fig. 38). He also inserted tubes (i.e. created fistulae) into the cardiac portion of the stomach and into the duodenum. Through these fistulae he was able to place meat extract directly into the gut. He was interested in discovering where he had to place meat in order to stimulate the secretion of gastric juice. The work involved two groups of dogs, a test group and a control group and they were surgically prepared as shown in fig. 39. The difference between the test and control group was that in the test group the stomach was divided distal to the pylorus, thus separating the pylorus from the duodenum, whereas in the control group the stomach was divided in such a way that the pylorus retained its connection with the duodenum but was separated from the rest of the stomach.

The results of the experiment were as follows:

Position of the food	Response of the test	Response of the control
Meat extract in stomach	Pouch secreted	Pouch did not secrete
Meat extract in duodenum	Pouch did not secrete	Pouch secreted

From these results Pavlov concluded that the pylorus was the sensitive area of the stomach which initiated a reflex action which resulted in the secretion of gastric juice by the stomach pouch. At this time, in 1895, Pavlov was engaged in work on salivary and gastric secretion and he had shown that nervous mechanisms were involved. In interpreting the results of his experiments with the stomach pouch, Pavlov

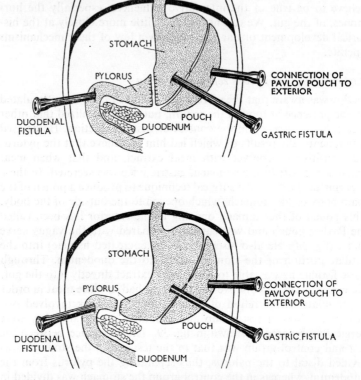

FIGURE 39. The surgical preparation used by Pavlov in 1895 to demonstrate that the pylorus is the area of the stomach which is sensitive to the presence of food and is capable of initiating the flow of gastric juice from the fundus.

assumed that the connection between the pylorus and the gastric glands was a nervous one, although there was no evidence to indicate either a nervous or humoral reflex. In 1905 J. S. Edkins provided evidence that a hormone, gastrin, was manufactured by the pyloric antrum. No further attention seems to have been paid to the idea that a nervous reflex may be involved in addition to the clearly established humoral mechanism.

Edkins suggests a humoral reflex mechanism

The following note appeared in *Proceedings of the Royal Society of London*, Vol. 76, Series B in 1905. On the chemical mechanism of gastric secretion by J. S. Edkins, May, 1905.

'It has long been known that the introduction of certain substances into the stomach provoke a secretion of gastric juice. This is regarded as in no sense depending upon mere mechanical stimulation of the mucous membrane, and it has been thought that the nervous mechanism of the gastric glands may be susceptible to certain local chemical stimuli.

'On the analogy of what has been held to be the mechanism at work in the secretion of pancreatic juice by Bayliss and Starling, it is probable that, in the process of absorption of digested food in the stomach, a substance may be separated from the cells of the mucous membrane which, passing into the blood or lymph, later stimulates the secretory cells of the stomach to functional activity. The following observations support this view:

'If an extract in 5% dextrin of the fundus mucous membrane be injected into the jugular vein, there is no evidence of secretion of gastric juice. If the extract be made with the pyloric mucous membrane, there is evidence of a small quantity of secretion. With dextrin by itself there is no secretion.

'Extracts of fundus mucous membrane in dextrose or maltose give no secretion; extracts of pyloric mucous membrane give marked secretion; dextrose or maltose alone bring about no secretion.

'If extracts be made with commercial peptone, it is found that no secretion occurs with the fundus mucous membrane, a marked secretion with the pyloric mucous membrane; the peptone alone gives a slight secretion.

'If the extracts be made by boiling the mucous membrane in the different media, the effect is just the same, that is to say, the active principle, which may be called "gastrin", is not destroyed by boiling.

'Finally, it may be pointed out that such absorption as occurs in the stomach apparently takes place in the pyloric end. With the pig's stomach, in which the true cardiac region differs from the typical fundus region in having only simple glands as in the pyloric, extracts of the cardiac region in general have the same efficacy in promoting secretion, as do pyloric.'

A long and confusing search went on for confirmation of Edkin's gastrin. Some of the confusion arose because extracts made from

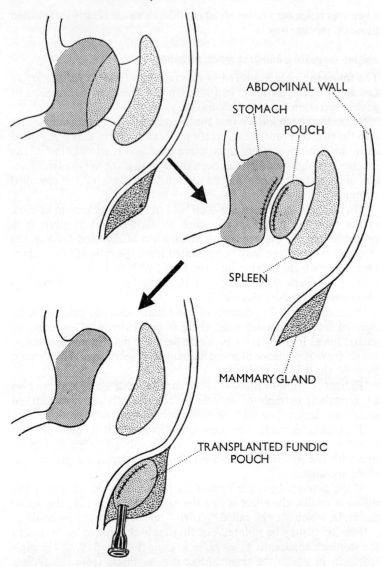

ABDOMINAL WALL

STOMACH

POUCH

SPLEEN

MAMMARY GLAND

TRANSPLANTED FUNDIC
POUCH

FIGURE 40.   The auto transplanted fundic pouch created by Ivy &
Farrell 1925. The 'vagally separated' Hedienhain pouch is transplanted
within the mammary tissue in the ventral abdominal wall. It is now
completely isolated from the influence of the vagal nerve. The pouch
secreted free acid when the dog was fed indicating that a humoral factor
was involved.

many tissues will cause secretion of acid by the stomach. Thus it was said that the pyloric hormone, gastrin, was not specific to this part of the gut and since it could be isolated from many tissues the hormone was not of physiological significance. Many years were to pass before histamine was recognized to be the non-specific substance, found in many organs, which caused secretion of stomach acid. It was not until 1938 that a protein was recognized in the pyloric extract which caused gastric secretion. Another reason for confusion about gastrin was that experiments designed to prove the existence of this hormone in the pyloric antrum necessitated the complete denervation of the stomach in order to exclude the possibility of any nervous activity being responsible. Unfortunately, in a completely denervated stomach, gastrin is not released from the pylorus and the glands of the fundus do not respond to gastrin, even if it is present. Another complicating factor is that gastrin is only released when the pH in the pylorus is more than 2·5. Experiments using pure gastric juice in the pylorus all failed because the gastric juice was far too acid to allow the hormone to be released. Much of the experimental work utilized Pavlov pouches and irrigation of the pylorus with fluids supposed to release gastrin, but by these means no one was able to stimulate the fundus to release significant amounts of gastric juice. By 1922 it seemed that there was no valid physiological evidence to support the gastrin theory. Experiments were performed involving crossing the blood systems of dogs and they failed to show any circulating hormone. Confirmation of the gastrin theory came eventually but it had to wait for a new surgical technique—the transplanted fundic pouch. In 1925 Ivy and Farell succeeded in completely transplanting a fundic pouch (fig. 40), but the first convincing evidence for the gastrin theory came from the experiments of Gregory and Ivy in 1941. They isolated the stomach and created a permanent opening to the outside after cutting the vagus nerve supply. The oesophagus was sewn on to the duodenum and a pouch of the stomach was transplanted into the body wall. When the dogs were fed via the mouth, the stomach pouch which contained the pylorus, secreted gastric juice and so did the fundic transplant. It was concluded that the secretion of the isolated fundus was in response to some hormonal gastric stimulant arising from the intestine, since this was the only part of the gut (except the oesophagus and ileum) which was in contact with the food. There was no vagal stimulation to the stomach since the vagus had been cut. In the case of the transplant there was certainly no nervous connection since all the nerves were severed during the transplantation and there was not even direct vascular connection with the rest of the gut. The transplant received its blood supply from new vessels which had grown out into it from the body wall following transplantation. Any connection between the intestine and the fundic pouch must have been via the general circulatory system.

Further experiments were carried out using this preparation by Gregory and Ivy. When they irrigated the stomach pouch containing the pyloric antrum with meat extract, the stomach pouch secreted gastric juice but so also did the fundic transplant. They concluded that either the gastric hormone came from the mucosa, or that there was some substance in the extract (a secretagogue) which was being absorbed by the pylorus and was causing the hormone to be released elsewhere. They then irrigated the stomach with the local anaesthetic procaine followed by liver extract and now the fundic transplant did not secrete gastric juice although the main stomach continued to respond and secrete gastric juice. They knew that procaine did not interfere with the absorption of substances from the stomach and therefore it seemed clear that the secretion of the isolated fundic transplant was in response to the liberation of a hormone from the stomach. These results gave the first really strong evidence that a gastric hormone did in fact exist. Later, other workers showed that the pylorus produced a hormone when it was distended, even in the absence of food and this ruled out the idea that secretion of the hormone resulted from the absorption of some secretagogue in the food.

The identification and synthesis of gastrin

By the late 1950s Harper had prepared crude extracts of the hormone gastrin from the pyloric antrum and claimed that it could stimulate the secretion of acid juice from the fundus of the stomach when injected intravenously into anaesthetized cats.

Gregory injected this extract intravenously into conscious dogs and to his disappointment—and that of Harper—he could elicit no secretion of acid juice from the stomach. If the extract had caused acid secretion perhaps there would have been little further investigation, but in the presence of these conflicting results Gregory and his colleagues looked more closely into the problem of the extraction of gastrin. When Harper in 1961 published his paper describing his method for the extraction of gastrin, Gregory repeated the work and found that when the extract was injected subcutaneously into conscious dogs it proved to be an excellent stimulant for the secretion of acid juice from the stomach. The reason for his initial failure to obtain a response from conscious dogs was due to the mode of administration of the extract. Grossman in 1963 showed conclusively that if gastrin extracts are given by rapid intravenous injection they strongly inhibit the secretion of acid juice by the stomach in conscious animals, whereas if they are administered subcutaneously the extracts stimulate acid secretion. There must have been many workers who during the 1950s concluded falsely that extracts from the pyloric antrum contained no gastrin because the extract had been administered by the wrong route.

The method for extraction of gastrin which was published by Harper in 1961 was very similar to Edkin's original method of 1906 except that Harper had removed virtually all the histamine by precipitation using acetone. This work provided the first clear proof that Edkin's extract really did contain gastrin as well as histamine.

Harper's method of extraction was only suitable for small-scale extraction, but Gregory in 1962 evolved a method for large-scale extraction. He found that some cellulose derivatives, e.g. aminoethyl cellulose would take up gastrin from dilute solutions at low pH and low salt concentrations and would release gastrin when the pH and salt concentration was high. Sephadex columns, countercurrent distribution systems and chromatographic techniques were employed to give a gastrin which workers believed was a reasonably pure substance. Its molecular weight and amino-acid composition was worked out and the results of high-voltage electrophoresis and end-group determinations indicated that gastrin was a small peptide.

Soon improvements in the methods allowed 600 antrums to be processed weekly at Liverpool University (despite the competition for the stomachs from the tripe industry!). On Christmas day 1962 Gregory found that the extraction of gastrin from the first batch of 600 antrums had produced two slightly different gastrins which he called gastrin 1 and 2. By an extra chromatographic separation using an aminoethylcellulose column both gastrins were obtained in pure form.

Gastrin 1 and 2 had identical amino-acid composition which was as follows:

Aspartic acid 1.
Glutamic acid 6.
Glycine 2.
Alanine 1.
Methionine 2.
Proline 1.
Phenylalanine 1.
Tyrosine 1.
Tryptophane 1.
Ammonia 1.

The molecular weight was calculated as 2,114.

By 1964 the sequence of amino acids in the gastrin molecule had been determined using traditional methods of enzyme and acid hydrolysis, stepwise degradation of peptides and end-group determination. The task of synthesizing the molecule of gastrin was undertaken by the chemists. The molecule was made in three parts which were then joined together. A five-amino-acid peptide was joined on to an eight-amino-acid peptide and then a four-amino-acid peptide was added (fig. 41). The difference between gastrin 1 and 2 is that the latter is more acidic because the tyrosine of the molecule is present as tyrosine O-sulphate. By removal of the sulphate gastrin 2

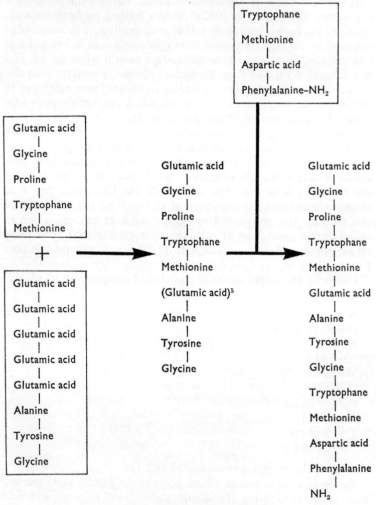

FIGURE 41. The synthesis of gastrin.

can be converted into the less acidic gastrin 1. The significance of this minute difference is not known.

SUMMARY OF THE PRESENT KNOWLEDGE ABOUT GASTRIN

We now know that the pyloric antrum of the stomach secretes a hormone which controls the production of gastric juice. Stimuli which cause the secretion of this antral hormone include local chemical and mechanical influences. The antral hormone, gastrin, causes the fundic glands to secrete acid and water but little pepsin. In the release of the hormone from the mucosa of the pyloric antrum into the blood stream there is some local nervous mechanism in which the vagus nerve most probably plays a part. Free acid in the pylorus inhibits release of the hormone. The range of physiological actions of gastrin is large, in addition to its effect on the fundus it causes strong stimulation of pancreatic enzyme flow and a weak stimulation of the flow of bile and bicarbonate from the bile duct.

Secretin and the pancreas

As early as 1825 Leuret and Lassaigne had noticed that the presence of acid in the duodenum stimulated the secretion of pancreatic juice. This work was, however, overlooked, and it was not until Dolonski, one of Pavlov's pupils, in 1894 discovered the copious secretion of pancreatic juice when dilute acid was put into the duodenum, that the story of the unravelling of the factors controlling the nature and timing of the secretion of the pancreatic juice really began.

Dolonski noticed that when dilute acid was placed in the rectum no secretion of the pancreas resulted. He assumed that the acid could be equally well absorbed by the rectum as the duodenum and concluded that from either source the acid must enter the circulating blood. Why, then, did acid in the duodenum stimulate pancreatic secretion, while acid in the rectum did not? Dolonski and Pavlov concluded that there must be some nervous reflex mechanism present in the duodenum but not in the rectum. However, further attempts to find a nervous pathway failed.

In 1901 Popeilski found that acid in the duodenum could cause the pancreas to secrete even when the splanchnic (sympathetic), vagus (parasympathetic) and the coeliac sympathetic ganglion had been cut, and even after the medulla of the brain and the spinal cord had been destroyed. Acid also stimulated secretion of the pancreatic juice even when the pyloric stomach was separated from the duodenum, provided that acid was put into the duodenum. It was believed that all the nerves to the pancreas came via stomach pathways and this posed the question of how a nervous reflex mechanism could be involved when the nerves supplying the pancreas had been severed so effectively. It was concluded that the nerves involved in the reflex must have a very short pathway between the wall of the duodenum and the pancreas. These were the days when thought was being concentrated on nervous reflex mechanisms.

Wertheimer and Lepage then showed that although the drug atropine blocked the action of the parasympathetic vagus nerve on the pancreas it did not effect the response of the pancreas to acid in the duodenum. Therefore in 1901 the general opinion was that secretion of the pancreas was controlled by a sympathetic nervous reflex.

On 16th January 1902 Bayliss and Starling performed their now famous and classical experiment which showed that a non-nervous reflex was involved.

The classical experiment of Bayliss and Starling

The experiment was performed on a dog in which the jejunum was cut away from the duodenum and ileum and all the nerves to the jejunum were cut, so that the jejunum was only attached to the body by the blood vessels. The cut end of the duodenum was sewn on to the cut end of the ileum. Using this animal it could be shown that when dilute acid was pipetted into the isolated jejunum the pancreas responded by secretion of its digestive juice. Since the only contact between the jejunum and the rest of the animal was by way of the blood stream Starling concluded that there must be a chemical reflex involved, substances being made in the wall of the jejunum which passed via the blood stream to the pancreas, causing it to secrete. It is reported that as soon as Starling saw the pancreatic juice being secreted as a response to acid in the jejunum, he rapidly cut off a piece of the jejunum and rubbed its mucous membrane with sand and dilute hydrochloric acid. He filtered the suspension and injected some of the filtrate into the jugular vein of the dog. After a few moments the pancreas responded by a greater secretion than ever before. Starling is reported as having said, that, 'It had been a great afternoon'. They named the substance, which was formed when acid acted upon the jejunal mucosa, 'secretin'. Later work showed that the acid was not the humoral agent itself, nor was the hormone made in the blood stream. It was made in the mucosa when the mucosa was stimulated by acid. Intravenous injection of acid extracts, from parts of the gut other than the duodenum and jejunum, e.g. from the oesophagus, rectum and stomach, caused very little secretion of the pancreatic juice when compared to a duodenal extract. The small secretory response of these other organ extracts is now known to be due to the presence of histamine, although at the time it caused some controversy and it was argued that since secretin could be extracted from a wide variety of organs the hormone was unlikely to be of much physiological significance in the living animal in the regulation of pancreatic secretion. We now know that the largest amounts of secretin are produced by the duodenum and the upper two-thirds of the small intestine. Little secretin is produced by the ileum.

Control of the pancreas by nerves (Pavlov) or by secretin (Starling)

By the late twenties Pavlov had perfected an experiment on an anaesthetized dog, which he had developed to such a fine art that he could perform it as a demonstration before lecture audiences. The dog had been previously operated upon, so that the pancreatic duct opened to the outside of the body and the rate of discharge of the secretion could be measured, that is the dog had a permanent pancreatic fistula. Pavlov anaesthetized the dog and then by electrical stimulation of the vagus nerve in the neck, using an induction coil, he was able to show that pancreatic juice flowed after a latent period of about 2 minutes following the electrical stimulation of the nerve, and that the flow of juice continued for about 5 minutes after the electrical stimulation. To Pavlov it appeared, 'to be definitely settled that the vagus nerve is the secretory nerve of the pancreas'.

When Bayliss and Starling tried to repeat Pavlov's experiments they were unable to do so and they therefore came to the conclusion that only a humoral agent caused the secretion of the pancreatic juice. One of Pavlov's pupils (Anrep) visiting London in 1912 demonstrated convincingly to Bayliss and Starling how Pavlov performed his experiment. Bayliss and Starling had failed to repeat successfully Pavlov's experiment because they had injected morphine into the dogs to sedate them before giving them a general anaesthetic. Unfortunately one of the side effects of morphine is that it causes spasm of the pancreatic duct so preventing the outpouring of secretions from the gland.

When the experiment of Bayliss and Starling was demonstrated to Pavlov the effect of secretin was self-evident. Pavlov is reported to have watched in silence and then to have disappeared into his study for half an hour. When he returned he said, 'Of course they are right, it is clear that we did not take out an exclusive patent for the discovery of the truth.' From this point onwards it was clear that the control of the pancreas was not by one system alone but that both humoral and nervous mechanisms were involved.

Pancreozymin and pancreatic juice

For many years there was disagreement between various workers as to whether secretin had any effect on the discharge of enzymes into the pancreatic juice. After the work of Bayliss and Starling all agreed that secretin extracted from the duodenal mucosa could be used to cause a flow of pancreatic juice, but people disagreed about the nature and volume of the juice. Some workers claimed that the juice had little amylase activity and that secretin stimulated a flow of water and salts only, while others claimed the juice to be viscid and of high amylase activity.

During the 1940s Harper and his co-workers realized that the confusion arose because of the different methods of extracting the hormone secretin from the duodenal mucosa. There were two main ways of extracting the hormone.

Method I	Method II
Extract mucosa with acid, precipitate using common salt, wash the putty-like precipitate with alcohol then re-precipitate with trichloroacetic acid	Extract mucosa with alcohol, precipitate using bile salts, wash the precipitate with alcohol
Effect	*Effect*
Pancreatic juice contains enzymes	No enzymes in the pancreatic juice
Active hormones present	*Active hormones present*
Secretin	Secretin only
Pancreozymin	

It became apparent that in method II, secretin was precipitated by the bile salts and that pancreozymin had been discarded in the solution. It was now accepted that there were two hormones released from the duodenal mucosa when acid was present in the lumen. They were:

1. Pancreozymin which stimulated the pancreas to release enzymes.
2. Secretin proper, which stimulated a flow of water and salts from the pancreas, but which did not stimulate the flow of enzymes.

The early workers, using crude secretin extract, had been using a mixture of the two hormones.

Summary of factors controlling the flow of pancreatic juice

We have seen that since the beginning of the 20th century opinions about the control of the flow of pancreatic juice have changed as new evidence has accumulated and as the prevailing fashionable ideas have fluctuated. At the beginning of the century, Pavlov's work on nervous reflexes dominated the scene. The vagus nerve was thought to be in sole control. Bayliss and Starling were convinced by their work that chemical reflexes were the controlling agents. We have seen how the controversy was resolved. The major advances of the fifties was in the recognition of the functional integration of nervous and humoral mechanisms in the digestive tract. There are three major factors governing the output of pancreatic juice:

1. Parasympathetic nerve (the vagus) produces enzyme-rich juice.
2. Pancreozymin produces enzyme-rich juice.
3. Secretin produces water and salts but no enzyme.

Figs. 37A, B, C show the kind of data which serves as evidence for this present viewpoint and they also show the nervous and humoral systems interact. In all the three cases a regular injection of secretin into the anaesthetized cat produced a steady flow of juice. With secretin alone there was little amylase activity in the juice. The effect of injecting pancreozymin or of stimulating the vagus is very similar, resulting in an increased amount of amylase in the juice without increasing the volume of the juice. The action of the vagus is very similar to pancreozymin except that atropine will block the effect of vagal stimulation whereas it does not alter the action of pancreozymin. This indicates that the pancreozymin is not having its effect via the vagus.

Substances affecting the release of secretin and pancreozymin

Acid in the duodenum is the best-known stimulant of duodenal hormones, but it is not the only one, for we know that removal of the stomach or a reduction in the acidity of gastric juice need not affect digestion. In experiments on dogs it has been shown that even if the duodenal contents are neutralized by bicarbonate, after a meal of meat the pancreas will still produce its secretion.

The effectiveness of differing duodenal contents on the secretion of hormones by the duodenum

Material in the duodenum	Effects on secretion of pancreozymin	Effects on secretion of secretin
Acid	Weak	Powerful
Amino acid and peptone	Very strong	Fairly powerful
Fats and soaps	Fair	Fair
Carbohydrate	Poor	Poor

Thus, it is seen that the products of digestion from the stomach are effective in releasing the flow of pancreatic juice by stimulating the duodenal mucosa to secrete hormones.

The flow of pancreatic juice in the normal digestion of a meal

The relationship between pancreatic secretion, duodenal contents and the rate of emptying of the stomach may be thought of as self-regulatory mechanism. In a fasting dog the flow of pancreatic juice is low but is periodic, possibly due to regular infrequent emptying of the stomach which pours gastric juice and saliva into the duodenum. When the dog is given a meal, secretion of the pancreas begins within a couple of minutes and the juice is soon flowing rapidly. Perhaps the initial slow rate of flow is controlled by the vagus reflex and then as soon as the acid and peptone containing stomach contents are released into the duodenum, secretin will be released and there will be a rapid flow of pancreatic juice. The continuous flow of pancreatic enzymes during the later stages of a meal is perhaps mainly due to the release of pancreozymin from the duodenum.

14

Hormones and Reproduction

Introduction

The sexual reproductive process in mammals involves copulation of the two sexes and during this process the sperm is placed inside the female reproductive tract at a time when the eggs are about to enter it. In order for egg and sperm to be present at the same time a great deal of co-ordinated activity is necessary. The sperm and egg must have been brought to maturity within the male and the female and the female reproductive tract must be prepared to receive the egg and nurture it during its subsequent development *in utero*. The behaviour patterns of the two sexes must be such that the female will permit the male to inseminate her. The young must be born at a time when there is, in the wild state, appropriate food for the young to eat after they have been weaned. Thus the breeding season must be timed to suit the climatic conditions. Milk must be available from the mother at the time of birth, and mammary glands must be prepared for this rôle during the pregnancy. If any one of these processes does not occur at the appropriate time in the sequence of reproductive activity, then the animal will be at a disadvantage in the competitive business of life. Evolutionary success depends on the ability to produce offspring which will grow healthily to maturity. The reproductive processes in male and female mammals are regulated by external factors in the environment and by internal factors within the organism. We can now look in some detail at the way in which hormones play a part in reproductive activities.

Hormones in the male

The reproductive structures in the male may be classified into two groups: the primary sex organs, the testes and the secondary sex organs. The secondary sex organs include genital ducts and glands, e.g. vas deferens, penis, seminal vesicles and prostrate gland, and non-genital features which distinguish the sexes, e.g. body size and shape, hair distribution, voice.

The testis consists of a mass of seminiferous tubules which produce the sperm. The seminiferous tubules (plate 3b) are lined by the seminiferous epithelium consisting of two types of cells, the germinal cells and the sustentacular cells of Sertoli. The majority of cells in the seminiferous epithelium are germ cells in various stages of development towards mature spermatozoa. Sertoli cells serve as points of attachment for the spermatids during their maturation into spermatozoa. In between the tubules is the insterstitial tissue of the testis, consisting of cells of an irregular polyhedral shape, the cytoplasm of which contains numerous mitochondria, a Golgi apparatus and granules composed of lipids (plate 3b).

The testis produces the male sex hormone or androgen, which regulates the growth and maintenance of secondary sex characters. The first observations which implicated the interstitial tissue of the testis as the source of androgen, was provided by Ancel and Bouin (1903). They recognized that in animals in which the testes do not undergo their normal descent from the abdomen into the external scotal saca, i.e. cryptorchid animals, the normal male secondary sex characters are still retained in spite of the fact that the seminiferous epithelium of the tubules of the testes were found to be infantile in character. By ligating the efferent ducts of the testes of guinea-pigs and rats, Ancel and Bouin caused a complete degeneration of the seminerous epithelium, while the interstitial tissue remained normal. After this treatment the male secondary sex characters remained normal. The seminiferous epithelium can also be destroyed by exposure to X-rays. The interstitial tissue is much more resistant to X-rays and remains histologically normal. After exposure to X-rays the secondary sex characters remain intact and may even be larger than those of normal animals. Further evidence supporting the view that interstitial tissue is the source of testicular androgen comes from the observations of Pézard, who found that extracts of the cryptorchid testis of swine could stimulate the growth of the capon's comb—a characteristic effect of androgens. The testes of the cryptorchid swine contained an infantile seminiferous epithelium but normal interstitial tissue. Experimental techniques which cause preferential degeneration of interstitial tissue have also been performed. Deficiencies of the vitamin B complex in the rat cause an involution of the seminal vesicles and prostrate gland. The testes of these rats show degenerate interstitial tissue but a normal seminiferous epithelium producing spermatozoa.

The testis and pituitary gland

When the pituitary gland is removed from the male animal, the testes atrophy (involving both seminiferous epithelium and interstitial tissue) and there is a progressive involution of the secondary sex characters. These effects of hypophysectomy can be prevented or reversed by the injection into the animal of extracts of the pituitary glands of other animals, or by transplantation of pituitary tissue. The anterior lobe of the pituitary gland produces three gonadotrophic hormones: follicle-stimulating hormone, FSH, luteinizing hormone, LH (or interstitial cell-stimulating hormone, ICSH) and luteotrophic hormone (prolactin). Luteotrophic hormone is concerned, in the female, with the regulation of the activity of the corpus luteum and with lactation. It will not be considered further in this section. The two hormones FSH and LH are not sex specific. In the male FSH causes the development of the seminiferous tubules and the maintenance of spermatogenesis, and in the female it controls the development of the ovarian follicles up to the point of ovulation. The two names, LH and ICSH refer to the same pituitary hormone. It was once thought that these were two hormones, one causing luteinization of the ruptured ovarian follicle, the other regulating the activity of the interstitial tissue of the testis, but these actions are now regarded as two effects of a single hormone.

The rôle of the pituitary gland in the regulation of gonad function was indicated by the early observations of Smith and Engle (1927) and Ascheim and Zondek (1928) who found that the pituitary glands of several species contained substances which when injected into sexually immature animals could stimulate the development of the gonads and provoke the appearance of precocious sexual maturity. The establishment of the existence of two distinct pituitary hormones regulating the activity of the gonads results from many kinds of experimental study. Some studies have used methods for the extraction of active material with predominantly FSH or LH activity. The urine of pregnant women has a gonadotrophic activity which is characteristic of LH rather than FSH, whereas the urine of women from whom the ovaries have been removed, or the urine of normal men, has a gonadotrophic activity which is characteristic of FSH. From these various natural sources active material can be extracted which has almost pure LH or FSH activity. Extracts of pituitary glands contain both FSH and LH, but the two substances can be separated by appropriate techniques.

Parabiosis and the investigation of pituitary gonad relations

One experimental technique which has been widely used in the study

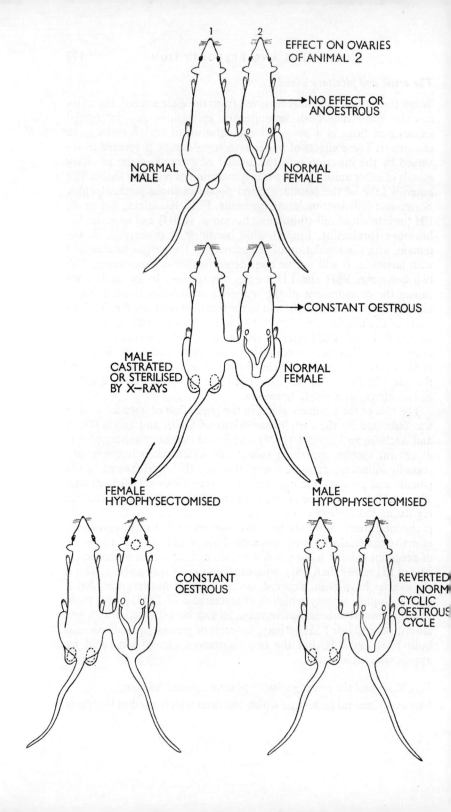

EFFECT ON OVARIES
OF ANIMAL 2

1 2

NORMAL MALE   NORMAL FEMALE   → NO EFFECT OR ANOESTROUS

MALE CASTRATED OR STERILISED BY X-RAYS   NORMAL FEMALE   → CONSTANT OESTROUS

FEMALE HYPOPHYSECTOMISED   CONSTANT OESTROUS

MALE HYPOPHYSECTOMISED   REVERTED NORM CYCLIC OESTROUS CYCLE of pituitary-gonad relationships is that of parabiosis (p. 23), and some examples of the uses of this technique are considered below. In 1930 Martins noted that if a normal female rat was joined to a normal male rat the oestrous cycle of the female partner was either unaffected or a state of anoestrus developed, i.e. there was no sexual rhythm. If a normal female was joined to a castrated male, or a cryptorchid male, this caused a state of constant oestrus in the female with excessive development of Graafian follicles but without corpus luteum formation. These results suggest that castration or cryptorchidism caused the pituitary gland of the male to produce an increased and continuous supply of FSH accompanied by little or no LH. In this type of experiment the female partner is used to indicate the predominant type of gonadotrophin produced by the male (fig. 42). This particular experiment also indicates the reciprocal nature of pituitary-gonad relationships. When the testicular source of androgen is removed, then the pituitary responds by an increased outpouring of FSH into the circulation. This suggests that usually testicular androgen exerts a restraining effect on the secretion of FSH from the pituitary gland. In 1932 other parabiotic experiments were carried out by Witschi and his co-workers in which male rats were sterilized by exposure to X-rays, thus leaving the interstitial tissue intact and the secondary sex characters maintained. When these rats were joined to normal females they produced a constant state of oestrus in the female indicating an excessive supply of FSH with little or no LH from the male partner (fig. 42). When the female of the parabiotic pair was hypophysectomized the constant oestrus persisted, indicating that the male was providing the supply of FSH. When the pair were separated and the female thereby removed from the source of gonadotrophin, a state of anoestrus developed in the female and all the mature Graafian follicles in the ovary underwent degeneration. If the pituitary gland was removed from the male of the partnership of a normal female and an X-ray sterilized male, then the state of constant oestrus in the female reverted to a normal oestrous cycle. Further proof that FSH production increases after castration was provided in 1935 when Du Shane et al. joined a hypophysectomized female rat with a normal male or female partner. The normal animal was unable to provide sufficient gonadotrophin to maintain the ovaries of the hypophysectomized female at the normal size. However, if the normal partner of the parabiotic pair was spayed, then the hypophysectomized female went into constant oestrus with large ovaries containing mature follicles but no corpora lutea.

FIGURE 42. Diagrams showing the use of parabiosis experiments in the study of brain–gonad relationships in the rat.

Testicular 'inhibin' and the pituitary

We have described how the technique of parabiosis has demonstrated that in the cryptorchid animal there is an increased rate of production of pituitary FSH. In the cryptorchid state the temperature of the abdominal cavity is inimical to spermatogenesis and the seminiferous epithelium returns to an infantile state. However, the interstitial tissue appears normal and some secondary sex characters, e.g. the prostrate gland may increase in size indicating an increased level of circulating androgen. In the cryptorchid animal, then, the increased output of FSH from the pituitary gland does not appear to be the result of a decline in the level of circulating androgen. It has been suggested that the seminiferous tubules are normally the source of some factor called 'inhibin'—which may be an oestrogen—which exerts a restraining action on the secretion of FSH by the pituitary gland similar to that of androgen derived from the interstitial tissue. Cryptorchidism, by causing degeneration of the seminiferous tubules will thus remove one component of the restraining influence of the testis on the pituitary gland, resulting in an increased production of FSH.

The two pituitary hormones FSH and LH have synergistic effects when they are administered together. When FSH is administered alone to animals there is little or no effect on the interstitial tissue of the testis and androgen production. However, if FSH is administered with ICSH, then the output of androgen by the testis is greater than if ICSH is administered alone. The mechanism of this synergism is unknown. Paradoxically sperm production has been maintained in some hypophysectomized animals by the administration of the androgen testosterone, but the significance or mechanism of this effect is unknown.

Androgens

Androgen is a collective name for substances which mimic the effect of testosterone in stimulating the development and activity of the male secondary sex characters.

We now know that the principal site of production of androgen in the male is the testis. The effects of surgical removal of the testes—castration—have been known since ancient times. Thus the ox (a castrate) is a much more manageable animal than the bull, and the gelding (a castrate) is much less fiery than the stallion. In man, castration has been carried out for a variety of reasons. Until the 18th century castration was practised on boy sopranos in order to maintain their high-pitched voice into adulthood—for the benefits of the opera and church choirs.

Early experiments which indicated that the testis was the source of masculinizing hormone include those of Berthold (1849) who showed that the effects of castration in the fowl could be reversed by means of a graft of testicular tissue (p. 2). In 1899 Brown-Séquard described now classical experiments performed upon himself in which he administered subcutaneously extracts of the testes of dogs and guinea-pigs. He observed an increase in mental and physical vigour following this treatment, although he wondered if these effects were due to self-suggestion, an ever-present problem in the interpretation of the effects of treatment in man. Other pioneer work in this field include the experiments of Walker (1908) and Pézard (1911) who injected extracts of testes into hens or capons and found a marked increase in size of the comb and wattles and the development of pugnaciousness and crowing.

These early experiments using testicular transplants or the injection of testicular extracts were made before the chemical nature of the active principle was identified. In 1927 McGee was able to extract a relatively pure substance from the testes of bulls which was a potent androgen, causing masculinization in the capon with the development of comb and wattles and the appearance of aggressive behaviour. In 1931 an androgen called androsterone was isolated in pure form from the urine of men. In 1934 this androgen was synthesized from the sterol cholesterol, and in 1935 a much more potent androgen, testosterone, was isolated in crystalline form from testicular tissue. Soon this compound was synthesized from cholesterol. Testosterone is the chief androgen produced by the testis. This hormone is degraded in the liver into substances of less biological activity, e.g. androsterone, which are excreted into the urine. Other weak androgens excreted in the urine—collectively known as 17-keto steroids—are derived from the adrenal cortex.

The actions of androgen

The actions of androgen in the body are widespread and they affect many processes not directly related to reproduction. Further, androgens produce effects in the female similar to those in the male—with obvious exceptions where the female does not possess the appropriate organs. Here we can consider only briefly some of the effects of androgen.

GENITAL TRACT

Each area of the genital tract and its associated glands is regulated by testicular androgen. The growth and activity of the scrotum, seminal vesicles, prostate gland, penis, bulbo-urethral glands, etc., are determined by androgen. Androgen can also influence the female genital tract. The administration of androgens to the pregnant female disturbs the differentiation of the genital tract of female foetuses, producing masculinized sex organs. Even in the differentiated female reproductive tract androgens may produce some features of masculinization, e.g. hypertrophy of the clitoris, the female homologue of the penis.

NON-GENITAL STRUCTURES

Androgens also influence non-genital secondary sex characters such as hair distribution, voice, distribution of fat, muscle size, etc. The rôle of androgen in the sexual differentiation of the hypothalamus is described in chapter 12.

In addition to these effects on secondary sex structures androgens have been found to influence the kidney, adrenal gland, liver, pancreas, thyroid, thymus, salivary glands, skin pigmentation and blood flow, red blood cell formation. It is obvious that very many body tissues are capable of responding to androgen. Some tissues have developed this responsiveness to a degree in which major changes in tissue activity result from the application of the hormone in physiological amounts.

MODE OF ACTION OF ANDROGEN

The treatment of animals with androgen produces, like oestrogen, marked changes in the structure of the target tissues—in their size, blood flow, secretions, cytological structure and biochemical composition. Recent evidence suggests that some of the effects of androgen appear to be due to gene activation. In 1962 Liao and Williams-Ashman found that ribosomes from the prostate gland of the castrated rat synthesized protein at a higher rate after the animal had received testosterone. Since the addition of the hormone directly to the ribosomes isolated from the castrate rat did not also stimulate protein synthesis it seems that the hormone acts at a stage preceding ribosomal function. In 1964 Wicks and Kenney found that within an hour of administering testosterone to castrate rats there was an increase in the synthesis of RNA in the prostate gland. In one species tritiated testosterone has been found to become bound to nuclear chromatin (duck's preen gland).

On bone marrow testosterone has another effect. The action of testosterone here is to stimulate the production of red blood cells which involves multiplication of red cell precursors in the bone marrow. Autoradiographic studies of bone marrow incubated with testosterone and a radioactive precursor of DNA (tritiated thymidine) have shown that the hormone stimulates the formation and replication of DNA (Diamond, Jacobson and Sidman, 1967).

Reproduction and hormones in the female

Introduction

Like the testis the ovary serves two functions, the production of germ cells and hormones which regulate the activity of the secondary sex structures. In terms of endocrine function the ovary differs from the testis in that hormone production is not continuous, but fluctuates during the breeding period, and further, at least two hormones are produced by the ovary which are distinct in their structure and activity.

The development of the Graafian follicle and the corpus luteum

The ovary of mammals (plate 4a) has within its cortex Graffian follicles at all stages of development. Each follicle contains one potential ovum, but only about one in every thousand follicles matures. The infant human female has 400,000 oocytes in her ovaries and in 30 years of reproductive life about 400 of these may mature at a rate of about one per month. The follicles which are not to mature undergo a regression and are said to be atretic. The follicles which are to develop undergo an enlargement, the epithelium of the follicle proliferates and fluid (the liquor folliculi) accumulates within the follicle. The oocyte develops within the centre of the fluid-filled cavity of the follicle.

Eventually the oocyte is discharged from the follicle and it enters the oviduct and if it is fertilized it becomes embedded in the wall of the uterus to which it becomes attached by the placenta. The empty follicle becomes transformed into an endocrine gland—the corpus luteum. Some of the follicle cells enlarge and become filled with lipid droplets and pigment. The supporting connective tissue is rich in blood vessels. This important endocrine gland in women exists for about 14 days if the egg is not fertilized, but can last for 2–3 months if pregnancy ensues. The function of the corpus luteum is to prepare the mucous membrane lining the uterus to receive the ovum and its functional life is adequate for this purpose. At the end of its useful life the corpus luteum degenerates into a whitish body, the corpus albicans, which sinks deeper into the ovary and eventually disappears.

Ovarian hormones—1. Oestrogen

The term oestrogen is applied to a substance which produces cornification of the vaginal epithelium of the mouse similar to that which occurs during the phase of oestrus.

That the ovary regulates sexual functions by means of hormones was suggested by the observations of Marshall and Jolly (1905) who found that spayed female dogs could be brought into oestrus either by the transplantation of ovaries or the injection of ovarian extracts from a dog in oestrus. These workers were also able to postulate that the ovary produces two hormones, one hormone causing oestrus and

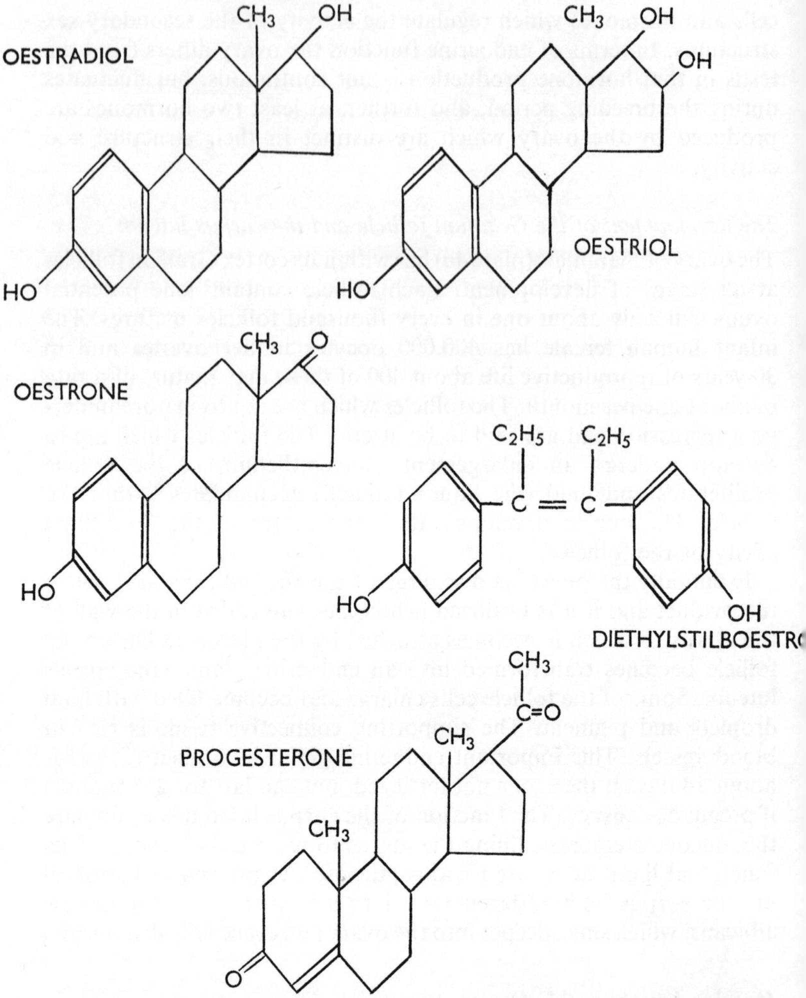

FIGURE 43. The chemical structure of oestrogens and progesterone.

the other produced by the corpus luteum. In 1923 Allen and Doisy found that the liquor folliculi, obtained from follicles of the ovary of the sow, could produce a cornification of the vaginal epithelium of the rat. The vaginal epithelium responds to the oestrogen in the circulating blood and it is possible to correlate vaginal structure with changes in the ovarian follicles throughout the oestrous cycle. The structure of the vaginal epithelium can easily be studied by scraping the epithelium with a moist swab on a stick, thus removing the superficial layers of the epithelium. Stained smears of these scrapings readily show that the vaginal epithelium has a characteristic appearance at each stage of the oestrous cycle. Only during the period of oestrus itself do cornified epithelial squames appear in the smear. This is the period when maximal oestrogen is present in the circulating blood.

The active principle present in ovarian extracts which causes vaginal cornification was first named 'oestrin' or 'folliculin'. In 1929 one active compound was isolated in pure crystalline form and was called oestrone (fig. 43). Later other pure substances, differing in chemical composition from oestrone, were extracted from the ovary. All of them produce similar effects on the vaginal epithelium, but they differ considerably in their potency, oestradiol (fig. 43) being the most potent compound. These various substances are known collectively as oestrogens.

There are now synthetic artificial oestrogen such as stilboestrol (fig. 43). These do not normally occur in animals but they have potent oestrogenic activity. They have a variety of uses, including the 'chemical sterilization' of the male; this effect is dependent upon the fact that like natural oestrogens they are potent inhibitors of gonadotrophin secretion by the pituitary gland and this can cause atrophy of the testis. This 'chemical sterilization' is put to practical use in the caponization of cockerels or in the treatment of carcinoma of the prostate gland in man. In the early days of stilboestrol manufacture many male factory workers became sexually impotent and developed enlargement of the breasts due to inhalation of stilboestrol.

Sources of oestrogens
The observations of Allen and Doisy showed that the follicular fluid has oestrogenic activity. However, even after aspiration of fluid from the large Graafian follicles, the ovary still contains considerable amounts of oestrogen. This observation led to the view that oestrogen is produced by the granulosa cells of the follicle which form a layer of cells adjacent to the follicular cavity. However, a variety of experimental observations showed that the granulosa cells of the ovary are not essential for oestrogen production. The granulosa cells are, for example, readily destroyed by exposure of the animal to X-rays and this procedure does not remove the source of ovarian oestrogen. In 1926 Ascheim studied the problem of which portions of the ovary could produce oestrogen by transplanting various parts of the human ovary into spayed mice. Changes in the vaginal epithelium of the mouse were used as an index of the amount of oestrogen in the circulating blood of the mouse. He found that neither granulosa tissue scraped from the wall of a large follicle or pieces of ovarian

G

cortex without follicles produced cornification of the vaginal epithelium when they were transplanted into a spayed mouse. However, when he transplanted ovarian tissue containing atretic follicles or thecal tissue (the outer layers of the Graafian follicle), or corpora lutea, or ovarian cortex removed during pregnancy, then the transplantation was followed by the appearance of cornification in the vaginal epithelium of the mouse. Thus several components of the ovary appear to be able to produce oestrogen, although the theca is probably the most important source of hormone.

Under normal conditions the ovary is the chief source of oestrogen. However, the urine of women from whom the ovaries have been removed still contains traces of oestrogen, perhaps derived from the adrenal cortex. The placenta also produces oestrogen in considerable amounts. Oestrogen is also produced by the male, principally from the testis. In 1934 Zondek found that the urine of the stallion contains very large amounts of oestrogen, from 10,000–400,000 m.u./litre, sometimes containing even more oestrogen than does the urine of a pregnant mare. He also found that two testes, weighing 350 g, removed from a stallion, contained 23,100 m.u. of oestrogen—this is the richest known source of oestrogen. After castration most of the urinary oestrogen of the gelding disappears. The oestrogen present in the urine of the stallion was identified as oestrone in 1935.

The actions of oestrogen

Like the male hormone, testosterone, oestrogens produce widespread effects in the organism. The actions of oestrogen on the hypothalamus affecting the secretion of pituitary gonadotrophins and sexual behaviour is discussed in chapter 12. Vaginal changes induced by oestrogen include cornification of the epithelium, hyperaemia, hypertrophy and oedema of the vaginal stroma, the appearance of glycogen in the epithelium, a fall in pH of vaginal secretions and a change in the electrical potential. The uterus is dependent upon a continuous supply of oestrogen for its growth and function. Oestrogen affects both the muscle of the uterus and the mucous membrane lining of the uterus. The rôle of oestrogen in determining the spontaneous activity and the sensitivity of the uterine smooth muscle to stimulants is described in chapter 4. Oestrogen is also involved, with other hormones, in the growth, maintenance and secretory activity of the mammary gland (p. 194).

In addition to the effects of oestrogen on secondary sex structures—uterus, vagina, external genitalia, mammary gland—and on the hypothalamus and pituitary gland, oestrogen also has marked influences on body growth and body weight and on tissues such as skin, adrenal gland, and on water and electrolyte metabolism. In most mammalian species the female is smaller than the male, and this is in part due to an effect of oestrogen in checking the growth of the female by an inhibitory effect on the secretion of growth hormone from the anterior pituitary gland.

Mode of action of oestrogen

The metabolic consequences of the administration of oestrogen are very varied. Even in a single cell, such as a smooth muscle cell of the uterus, the effects may include a rise in the trans-membrane potential, an increased tendency to generate pacemaker potentials, an increased sensitivity to stimulants, an increased synthesis of contractile protein, an accumulation of energy-rich phosphate and changes in enzyme content and respiratory rate.

One of the first observable effects of the administration of oestrogen on the uterus is a dramatic increase in the rate of RNA synthesis. This effect occurs in less than 30 minutes after the administration of the hormone. Initially there is a rise in the amount of messenger and transfer RNA, followed by the appearance of ribosomal RNA. But even earlier than the appearance of RNA is the activation of the enzyme RNA–DNA polymerase. These effects on RNA synthesis ultimately lead to an increase in the various cellular proteins, the contractile protein actomyosin and enzymes. This sequence of events is what one could expect if oestrogen triggered off these various changes by a primary effect at the level of the gene, i.e. on RNA synthesis. Indeed, if actinomycin (p. 43) is administered to the animal prior to oestrogen treatment, then none of these effects on protein synthesis appear. If all the many effects of oestrogen are to be explained in a similar way, then we must postulate that oestrogen is activating a considerable number of genes. The problem remains of explaining how a single hormone, oestrogen, can activate a whole range of genes in a special sequence and how it can activate them to different degrees.

In different target organs a different set of genes will be activated. Thus in the hen, oestrogen has an action on the liver, causing it to produce those proteins which ultimately appear in the egg yolk. These proteins are phosvitin and lipovitellin. The protein phosvitin contains large amounts of the amino-acid serine and prominent among the types of transfer RNA which appear in the cells of the hen's liver after treatment with oestrogen is that transfer RNA which is involved in the inclusion of serine into the growing protein molecule on the ribosome. The liver of the cock bird does not normally produce these yolk proteins, but even he can be made to produce them if oestrogen is administered.

Many problems remain in the analysis of the action of oestrogen on the genetic material of the cell nucleus, not only how the hormone can activate specific genes but how different genes are preselected to react to oestrogen in the various target organs.

Ovarian hormones.—2. Progesterone

In 1929 Corner and Allen found that extracts of the corpora lutea of the ovary could maintain pregnancy in rabbits from which the ovaries had been removed. These extracts could also produce those changes in the uterine endometrium suitable for the implantation of the fertilized ovum. The active principle present in these extracts was called progesterone. In 1934 four groups of workers almost simultaneously reported the isolation and identification of progesterone (fig. 43). Because more than one compound has the same action as progesterone the generic name progestin is now used.

Sources of progestin

The corpora lutea of the ovary are the main source of progestin. During the oestrous cycle there is a gradual transition from a phase of oestrogen dominance to one of progesterone dominance. Oestrogen is the principal hormone secreted by the ovary during the period of development of the Graafian follicles up to the point of ovulation. As the corpus luteum develops from the ruptured Graafian follicle, so does the secretion of progesterone begin and it eventually comes to dominate the ovarian endocrine scene. If fertilization of the ova and implantation do not occur, then the corpora lutea gradually undergo atrophy and new Graafian follicles begin to develop. However, if pregnancy occurs, then the functional life of the corpus luteum is maintained by a pituitary-like gonadotrophin which is secreted by the placenta. Indeed, the placenta may contribute directly to progesterone production and in some species the placenta is the most important source of progesterone during pregnancy. In addition to the ovary and the placenta, the adrenal cortex is also a source of progestin.

Unlike oestrogen and androgen there is little storage of progestins in the body and constant production has to keep pace with inactivation and excretion of progestin.

The biological actions of progestin

The special rôle of progestin is in the preparation of the endometrium of the uterus for the implantation of the fertilized ovum, and the modification of the activity of the muscular layers to produce a quiescent uterus which is conducive to the retention of the developing foetus. This function of progesterone is dealt with in some detail in chapter 4. Progestin is also concerned in the development of the mammary glands for lactation (p. 193).

Pituitary–ovarian relationships

The activity of the hypothalamus–pituitary complex in the secretion of the gonadotrophic hormones is discussed in chapter 12, and the regulation of testicular function by gonadotrophins is discussed on page 177. Here we shall be concerned only with the pituitary–ovarian relations which have not been discussed elsewhere.

The removal of the pituitary gland in the female results in the atrophy of the ovaries, especially the large follicles, so that the production of ova ceases. In addition the secretion of the sex steroids from the ovary declines and the various secondary sex structures—uterus, vagina, mammary glands—retrogress.

Three high molecular-weight proteins can be extracted from the anterior pituitary gland:

1. Follicle-stimulating hormone (FSH).
2. Luteinizing hormone (LH), which in the male is referred to as interstitial cell stimulating hormone (ICSH).
3. Luteotrophic hormone (LTH), which unfortunately has several synonyms, including prolactin and lactogenic hormone. It should not be confused with luteinizing hormone.

We can now consider the rôle of these hormones in regulating ovarian function.

FOLLICLE-STIMULATING HORMONE

In the female this hormone acts at the stage when the ovum is a large oocyte surrounded by several layers of granulosa cells. FSH causes a rapid swelling of the young Graafian follicle affected by the proliferation of the granulosa cells and the secretion of the liquor folliculi.

The very early growth of the follicles is independent of FSH or any gonadotrophin and it occurs even after hypophysectomy. The growth and maturation of the ovum, as distinct from the follicle, seems to be independent of FSH. The injection of extra FSH into an animal does not increase the rate of development of the follicle, but it does make more follicles develop and hence it increases the size of the ovary. The treatment of some infertile human females with FSH, extracted from human pituitary glands may thus result in super-foetation (multiple births).

ICSH

In the female this hormone allows the enlarged follicles to ovulate and to transform into corpora lutea, hence the name luteinizing hormone (LH) which is sometimes given to this hormone. Ovulation is usually ascribed to a sudden elevation in the amount of ICSH secreted by the pituitary. ICSH and FSH act in a very close and complementary fashion and probably one never acts without the influence of the other being important, i.e. they are synergistic.

LTH

Oestrogen and progesterone promote the growth of the mammary glands, but the actual secretion of the milk and its passage into the lumen of the ducts is under the control of the LTH. The final ejection of the milk is under the influence of another hormone called oxytocin, which is liberated from the posterior pituitary. The rat is the only animal, so far known, in which the hormone prolactin (LTH) also has a luteotrophic function, i.e. it will also prolong the functional life of the corpus luteum. In view of this fact it is perhaps unwise to consider LTH (prolactin) as a gonadotrophin. In fact, in many mammals and birds LTH can have an antigonadotrophic action and if it is injected into mature animals with functional gonads then these may even regress if the dose of prolactin is high and is continued for several days.

Pituitary activity during the oestrous cycle

FSH and ICSH are probably secreted by the anterior pituitary throughout the oestrous cycle, but the relative proportion in which they are secreted varies at the different stages in the cycle. In the early phases FSH is dominant and ICSH is less important. As the follicles get bigger they secrete greater amounts of oestrogen and the level of oestrogen in the blood rises, and as it does so it suppresses the amount of FSH secreted by the pituitary gland and brings about the secretion of ICSH so that at the point of ovulation there is much more ICSH than FSH.

Evidence from experiments on hypophysectomized rats showed that when these female rats were injected with FSH alone, the ovaries increased in weight but the uterus did not. Now the uterus is very sensitive to oestrogen and responds by rapid growth as oestrogen levels increase. Therefore in these experiments we must consider that pure FSH does allow the follicles to grow but not to secrete oestrogen. In the intact animal follicular growth is normally associated with increasing oestrogen production and uterine changes. It is known that if ICSH is administered as well as FSH then oestrogen is secreted by the developing follicle. This evidence suggests that the pituitary never secretes one gonadotrophin alone, but always the two in varying proportions.

The evidence that the amount of oestrogen produced by the ovary can regulate the amount of FSH secreted by the pituitary is very clear. Very minute amounts of oestrodiol, e.g. 0·015 μg daily, will lower the gonadotrophin output of the rat pituitary. Large quantities of oestrogen can cause complete inhibition of gonadotrophin secretion in males or females.

What part does oestrogen play in the release of ICSH by the pituitary? Experiments in rabbits, cows, sheep and rats have shown that oestrogen can cause ovulation. It does this presumably by causing an increase in the ICSH output, for this is the ovulatory

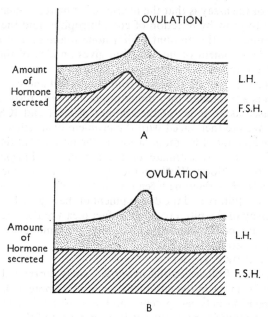

FIGURE 44. Hypotheses to explain the event of ovulation.

stimulating hormone. It could be that this is a most important rôle of oestrogen in the normal animal in that when oestrogen production reaches a certain level ovulation occurs. Experimental work on some animals indicates that progesterone is secreted by the follicle even before ovulation. We also know that progesterone can elicit the release of ICSH and cause ovulation.

There are two hypotheses which are commonly put forward to explain the data quoted above. They are shown in fig. 44 A and B. Note that in both hypothesis FSH and LH are always produced together. A is the classical explanation in which fluctuations of both FSH and LH secretion controls ovulation. B gives an alternative idea in which the amount of FSH is constant and only LH fluctuates, thus altering the balance of the two hormones. In neither view does the secretion of either hormone fall to zero at any point in the cycle. B is a simpler hypothesis for it does not call for the simultaneous control of both hormones but only for the control of ICSH. It is not a simple matter to decide between these two hypotheses at present. The amounts of circulating gonadotrophin are very small and they are assessed by bio-assay rather than by chemical measurement.

BIO-ASSAY OF GONADOTROPHIN

The basis of the assay is that the ovaries of immature females can be stimulated to grow by injection of gonadotrophins and the response is proportional to the amount of hormone injected. The degree of follicular development or luteinization gives an idea of the proportions of FSH and ICSH present.

If pituitary glands are removed from an animal and extracts from these glands are injected into immature females the response of the ovary is used as an index of the amount of FSH and ICSH in the pituitary. We are then faced with the problem of whether this assay reflects the amount of hormone stored in the pituitary at that time or whether it gives any estimate of the amount of hormone being secreted into the blood. Work on the pig showed a high coefficient of correlation ($r = + 0.69$) between the amount of gonadotrophic hormone in the pituitary and the development of the ovary. This suggests that bio-assay of the pituitary does in fact measure the rate at which hormones are being secreted rather than the amount of material merely being stored.

THE OVULATORY SPURT

The number of follicles in the ovary is closely correlated with the amount of hormone in the pituitary (the pituitary potency), but there is no correlation between potency and size of the follicle. In most mammals the follicles grow very little during most of the cycle except for a few days just before ovulation in the so-called ovulatory spurt. In this spurt the number of follicles is drastically reduced to the number destined to ovulate and these then grow rapidly. The reason for this spurt is obscure. If the rôle of oestrogen as a controller of FSH output is acceptable, then the spurt cannot be attributed to an increase in FSH output for as the follicles grow and begin to secrete oestrogen then surely the amount of FSH released from the anterior pituitary would fall. Perhaps one can assume that FSH production remains unchanged throughout the cycle and that many follicles regress because with increasing follicular size there is not sufficient FSH to maintain their growth. This idea is supported by the fact that the number of follicles actually ovulating can be increased by injecting FSH and LH. Also if the female has one ovary removed, then the remaining ovary will ovulate as many eggs as the two ovaries would have done. The amount of hormone available to the remaining ovary was twice what it would have been had two ovaries been present. The number of eggs produced by the ovary depends upon hormonal factors and not on any mechanism residing in the ovary itself.

The regulation of the development and function of mammary glands

The secretion of milk by the mammary glands is necessary for the young of most mammalian species during the early period of their adaptation to life outside of the uterus. The growth and function of the mammary glands is closely linked to other aspects of reproductive function so that the glands are prepared for secretion very shortly after parturition. A variety of endocrine mechanisms are involved in this synchronization of function, and the hormonal regulation of mammary function arising as it did late in phylogeny is somewhat more complex than the mechanisms controlling the reproductive tract itself.

Development and structure of the mammary glands

The glands are derived from modified skin glands during embryonic life. The number and position of the glands varies considerably from one species to another. During foetal life there is considerable development of mammary tissue, presumably under the influence of circulating hormones of the mother, and the breast of both male and female infants may sometimes be distended with secretion at birth. After birth the glands of the male involute. In the female the glands continue to develop, reaching full development towards the end of pregnancy, when the gland acquires secretory capacity.

In the human female before puberty the gland consists of rudimentary duct tissue extending a short distance from the nipple. At puberty, however, there is a sudden spurt of growth under the influence of hormones and the duct tissue elongates and branches producing a compound tubular gland. Each breast is composed of fifteen to twenty-five irregular lobes radiating from the nipple separated from one another by connective tissue, including abundant amounts of fat tissue (fig. 45A). Each lobe has an excretory duct—the lactiferous duct—which opens on to the nipple. In the breast of the non-pregnant female there is little glandular (alveolar) tissue.

During pregnancy marked changes appear in the structure of the gland. In the first half of pregnancy there is a rapid multiplication of the epithelium at the ends of the duct system and glandular secretory tissue appears. The interstitial connective tissue gradually disappears, making way for the great hypertrophy of epithelial and glandular

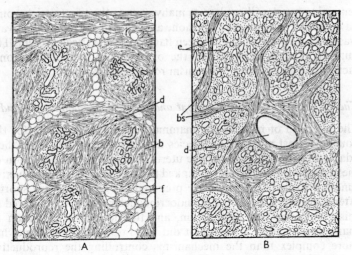

FIGURE 45. The structure of the human mammary gland. A shows a section through the mature, non-secretory gland. Here the lobules of the gland consist mainly of duct tissue which is embedded in dense connective tissue and fat cells. B shows terminal parts of the lobules of the lactating breast. The lobules contain masses of secretory tissue and are separated by sheets of connective tissue which contain the ducts.

components (fig. 45B). In the second half of pregnancy the glandular tissue begins to secrete. Initially the secretion consists of a clear glairy secretion called colostrum, but after the birth of the young this is rapidly replaced by milk.

After lactation has finished the gland regresses to its resting state, with disappearance of much of the secreting tissue. After cessation of ovarian function in old age the breast regresses even further, and comes to consist of a few scattered ducts embedded in connective tissue.

The effects of hormones on breast tissue

OESTROGEN

The administration of oestrogen is followed by some degree of breast development which varies in amount from species to species. In immature castrated guinea-pigs and monkeys oestrogen treatment produces development of duct and glandular tissue similar to that found in pregnancy. In the dog, however, oestrogen produces little effect on mammary growth. The administration of oestrogen to the virgin goat or heifer can cause lactation. In contrast, comparatively large doses of oestrogen can suppress lactation in some species. Even in a single species oestrogen may have different effects, depending upon the dose used and the state of the breast at the time of treatment.

PROGESTERONE

When administered alone in the prepubertal female progesterone may cause either duct or alveolar growth in some species. Much greater effects of oestrogen and progestin are observed when they are administered simultaneously and full mammary development under natural conditions occurs only when the two hormones are present.

The variable effect of sex steroids on breast development are in part due to the fact that the hormones influence the breast in several ways. First, they may act as direct stimulants of breast development. Secondly, they may exert indirect effects by way of the anterior pituitary, or at least they require the presence of pituitary secretions before their direct effect on the breast can develop. The removal of the anterior pituitary gland in most species largely prevents the response of the mammary gland to oestrogen or progesterone, and lactogenic hormone is only a partial replacement of whole pituitary action. Pituitary factors thus seem to be necessary for the full effects of the steroid sex hormones on the breast to be seen. In this the breast differs from the other secondary sex tissues, e.g. uterus and vagina which can respond fully to sex steroids in the absence of the anterior pituitary gland.

Lactation

Although the structure of the breast can be prepared for lactation by the actions of oestrogen and progesterone, a specific pituitary hormone—lactogenic hormone—is necessary for the secretion of milk. In the lactating animal the removal of the anterior pituitary gland results in an immediate cessation of lactation. In normal circumstances lactation begins very shortly after the birth of the young. This has been explained by assuming that during pregnancy oestrogen or progesterone inhibit the output of lactogenic hormone from the pituitary gland. Parturition results in an immediate decline in the level of circulating steroids, thus removing the inhibitory factors. However, the situation is complex and beset by the problem of species differences, and the effects of different amounts of hormones. Certainly large doses of oestrogen can inhibit lactation in the human female. Large doses of oestrogen are often administered to stop the flow of milk when it is decided not to breast-feed an infant. However, monkeys, goats and heifers subjected to prolonged treatment with oestrogen begin to lactate and may produce large amounts of milk.

In addition to the effect of prolactin and sex steroids the presence of other hormones such as thyroxine, growth hormone and corticosteroids are also necessary for normal lactation.

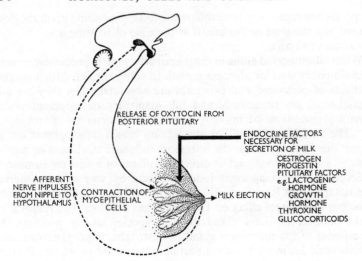

FIGURE 46. Summary of the factors involved in the secretion of milk
and in the ejection of milk from the mammary gland.

The hormonal regulation of milk ejection

The act of suckling produces a reflex ejection of milk from the nipple,
induced by the contraction of the myo-epithelial cells surrounding the
alveoli and smooth muscle of the ducts. The pathway of this reflex
involves the transmission of nervous impulses, arising in sense organs
on the nipple, to the CNS, and ultimately relay to the hypothalamic
centres which control the synthesis and release of the polypeptide
hormone oxytocin. This hormone released from the posterior pituitary gland stimulates contraction of the muscular elements of the
breast tissue, resulting in a rise in pressure of the duct system and
a discharge of milk from the nipple (fig. 46).

The act of suckling, in addition to causing the ejection of milk from
the breast, also seems to influence the actual secretion of milk in some
species. If suckling is permitted in some mammary glands in the rat,
then this stimulates milk production in other glands which are not
being suckled.

15

The Chemical Structure of Hormones in Relation to their Biological Activity

The chemical structure of many hormones has been elucidated, and increasing attention is being paid to the relationship between the chemical structure of hormones and their biological activity. In this study it is necessary to consider the physico-chemical properties which are concealed in the structural formula of the hormone molecule. Thus, for non-rigid molecules, many spatial orientations of the molecule are possible so that it is difficult to know in which molecular orientation the hormone meets and reacts with the target cell. It is possible to approach the problem of determining the biologically important parts of a hormone molecule in various ways, e.g. hormones which are large protein molecules can be partially degraded by treatment with proteolytic enzymes and the resulting simpler compounds can be tested for biological activity.

Another approach to the study of the relationship between physico-chemical properties of the hormone and biological effectiveness is to prepare 'isosteres' of the hormone which are then examined for their biological properties. In the preparation of an isostere part of the hormone molecule is substituted by atoms, molecules or ions, in which the peripheral layer of electrons can be considered to be identical to the component which has been replaced. In considering the suitability of isosteric groups for inclusion within a hormone molecule, it is important to consider the possible functions of this part of the molecule. An isostere which fulfils its functions by determining the spatial relationships of the hormone molecule, may not

197

fulfil those functions which depend upon electrostatic charges contributing to the formation of ionic or hydrogen bonds. Examples of isostoseres include the substitution of deuterium or fluorine for hydrogen atoms in a hormone molecule. However, in spite of the fact that deuterium and hydrogen atoms are very similar the isosteres can differ in their biological activity. Thus the deuterium bonds with carbon, oxygen and nitrogen are more stable than the corresponding hydrogen bonds with these elements. Fluorine isosteres may differ from the normal molecule in their stability in the body, because of the increased stability of the fluorine bond with carbon. Thus fluorinated thyroxine and fluorinated steroids may be more active than their non-fluorinated parent compounds.

Isosteres are rarely completely equivalent to the natural hormones. Many hormones have a multiplicity of actions and the substitution of isosteres which are appropriate for one biological function may not be appropriate for other functions.

Another useful technique is to produce a progressive chemical substitution of the hormone molecule, testing the effects of each substituent on biological activity. If one examines an homologous series of biologically active compounds it is often possible to observe a gradual change in activity with the change in chemical structure and the series may start as a biologically active compound and end with an inactive compound, or indeed one with different biological properties. The biologically inactive compounds may act as antagonists of the active members of the series. Thus dichloro-isopropyl-derivatives of noradrenaline have little noradrenaline or adrenaline-like activity and if they are applied to the biological object before the catechol amine they can block the actions of the hormone. Their chemical structure is such that the molecule of dichloro-isopropyl-noradrenaline will react and join with those active sites in the cell (receptors) to which the normal hormone becomes bound before it triggers off the biological effect. However, the structure of the isostere is such that it neither triggers off the biological effect nor does it leave the receptor at the same rate as does the normal hormone. Thus this derivative occupies and blocks the receptors and prevents access of the normal hormone.

Structure–activity relationships.—1. Gonadotrophic hormones

The bulk of human follicle stimulating hormone is protein in nature, but about one-fifth of the molecular weight is a polymer of several sugars. These sugars are members of a family of sugars called sialic acids. The simplest and perhaps most important of these sugars is N-acetyl neuraminic acid (NANA).

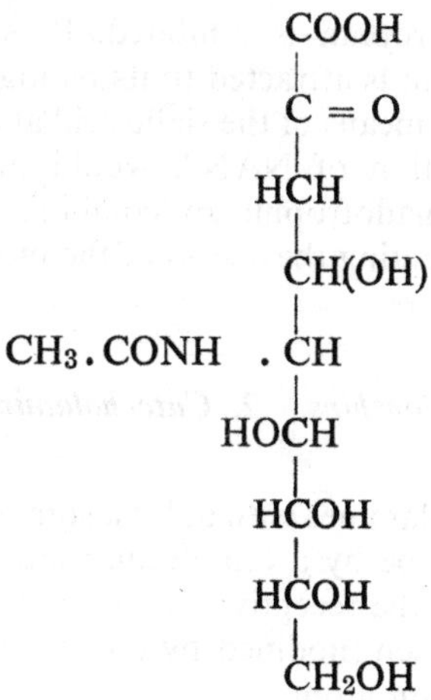

$$COOH$$
$$|$$
$$C = O$$
$$|$$
$$HCH$$
$$|$$
$$CH(OH)$$
$$|$$
$$CH_3.CONH \quad . CH$$
$$|$$
$$HOCH$$
$$|$$
$$HCOH$$
$$|$$
$$HCOH$$
$$|$$
$$CH_2OH$$

FIGURE 47. The structure of N-acetyl neuraminic acid.

Although NANA itself shows no gonadotrophic activity, if this sugar is removed from FSH, then the hormone becomes biologically ineffective. Thus NANA is essential for the biological activity of FSH. By contrast, some of the protein components of FSH can be removed, by the action of proteolytic enzymes, without considerable loss of biological activity. The view has been put forward that NANA is the key to the shape of the hormone molecule; the molecule of FSH contains only one negative charge, that of the acid radical of NANA, the rest of the molecule containing many positive charges on the side chains of the constituent amino acids. Presumably in this form the hormone is ensured effective transport to the target organ and protected from the FSH inactivation mechanisms which exist in the body. Some evidence for this latter function stems from observation that if FSH is mixed with NANA before it is injected into animals the biological effectiveness of the hormone is almost doubled. It is assumed that under normal conditions much of the FSH released from the anterior pituitary gland is rendered biologically ineffective by means of enzymes called neuraminidases which are present in various tissues which cleave NANA from the FSH molecule. These enzymes may be swamped by the administration of NANA to the animal so that administered FSH now reaches the ovary without much of it being rendered inactive.

The administration of NANA with chorionic gonadotrophin (derived from the placenta) has quite a different effect. Instead of potentiating the effect of the hormone as in the case of FSH, the action of the gonadotrophin is inhibited. It is possible that the chorionic gonadotrophin is attracted to its biologically reactive sites in the ovarian tissue by means of the sialic acid at the end of its sugar chain. The administration of NANA would thus antagonize the action of chorionic gonadotrophin by combining with the reactive sites in the ovary, preventing the access of the molecule of chorionic gonadotrophin.

Structure–activity relationships.—2. Catecholamines: adrenaline and noradrenaline

The reaction of catecholamines with at least some reactive sites in the target organs seems to be by a bimolecular attraction between the hormone molecule and the receptive site in the tissue. The molecule of noradrenaline has been modified by completely labelling it with deuterim on the α-carbon atom.

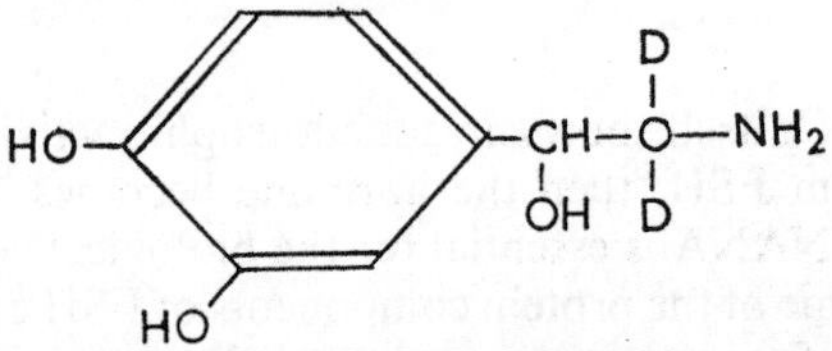

FIGURE 48. (—)-noradrenaline completely labelled with deuterium on the α-carbon atom.

This modified noradrenaline was tested on the nictitating membrane of the cat and found to be equally effective as the normal hormone. This type of evidence indicates that the molecule of noradrenaline triggers the excitatory response of the nictitating membrane by a bimolecular interaction of the hormone and the receptive site, involving electrostatic field effects and Van der Waals' forces, and not by a mechanism similar to that involved in the union of an enzyme and its substrate. The only absolute requirement for excitatory activity of the hormone is the presence of an amino group which should be preferably unsubstituted. The excitatory function of catecholamines can be shown with simple molecules such as 2-amino-heptate and this indicates that the other components of the hormone molecule may be concerned with other functions such as the attraction of the molecule to the receptive substance in the tissue. The excitatory effect of catecholamines are attributed to the cationic head of the molecule, which is thought to form an ion pair with an oppositely charged radical on the receptive substrate. When substitutions on the basic nitrogen of the molecule produce steric hindrance to the formation of ion pairs, then inhibitory rather than excitatory effects of the hormone appear.

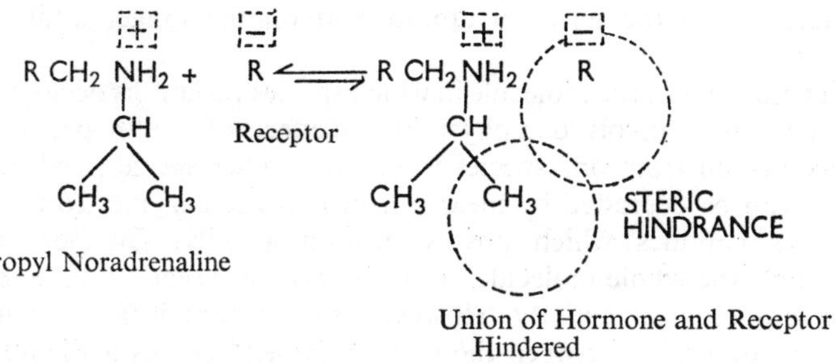

Isopropyl Noradrenaline

Union of Hormone and Receptor
Hindered

FIGURE 49. The steric hindrance to ion pair formation produced by substitution on the basic nitrogen of noradrenaline, as in isopropylnoradrenaline.

This suggests that the inhibitory effects of catecholamines (e.g. relaxation of smooth muscle) are dependent on the hydroxyl groups of the catechol ring rather than on the cationic head of the molecule. It is interesting to note that an isostere of isopropyl noradrenaline, viz. 3, 4-dichloro-isopropyl-noradrenaline, can prevent the inhibitory effects of catecholamines, presumable because it occupies the receptor sites and prevents the approach of the catecholamine phenolic groups to the receptor area.

Structure–activity relationships.—3. Insulin

The molecule of insulin consists of two chains of amino acids with the longer chain containing thirty amino acids and the shorter chain containing twenty-one amino acids. The two chains are linked

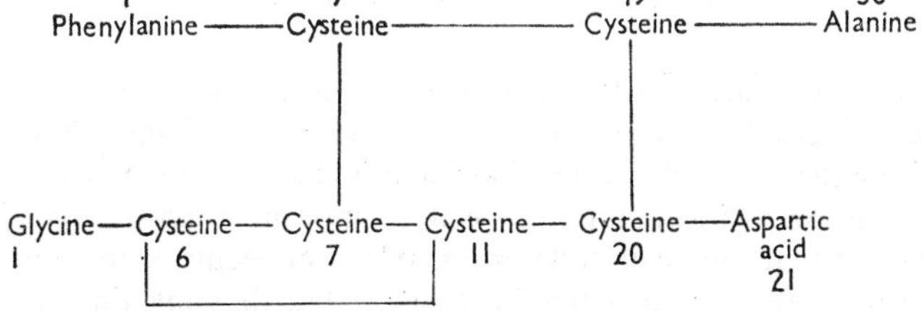

FIGURE 50. Simplified scheme of the insulin molecule. The numbers refer to the position of the amino acid in the chain.

together by the disulphide bridges of two cysteine residues. There is a further disulphide bridge in the shorter chain and it is within this part of the insulin molecule that major species variations occur in the amino-acid composition of the molecule. There are also species differences in the terminal amino acid on the longer amino-acid chain.

These areas of the molecule in which species variations occur, would seem not to be involved in physiological effects of insulin, particularly since insulin from one species is active in other species. Indeed, insulin can be degraded by means of proteolytic enzymes to produce smaller peptides which possess insulin activity. On both these accounts the whole molecule of insulin does not seem to be essential for physiological activity. However, insulin activity is lost if the disulphide bridge between the two amino-acid chains is disrupted.

The parts of the insulin molecule which are not concerned with the physiological effects of insulin at the cellular level may still have other important functions to perform. When insulin leaves the beta cells of the islets of Langerhan and enters the circulating blood it becomes bound to plasma proteins, in particular to several components of the beta-globulins. The liver is the first organ to be influenced by insulin liberated from the pancreas and here much of the insulin (up to 40 %) is bound, inactivated or destroyed. This is a protective mechanism which acts as a buffer for the potentially large amounts of insulin which can be released from the pancreas. It has been suggested that the formation of insulin–protein complexes in the blood plasma may protect the hormone from the various insulin-inactivating mechanisms of the body and allows a circulating store of the hormone to be formed. The existence of different kinds of carrier protein which react with and dissociate from insulin at differing rates could exercise a fine control on insulin release, and prevent the overwhelming effects on metabolism that the release of free pancreatic insulin might produce. Thus the physiologically inactive parts of the insulin molecule may react with plasma protein as a carrier for the essential part of the molecule.

To exert its physiological effects the hormone has to leave the blood. It seems that the hormone becomes 'fixed' at membrane sites. The brief exposure of a tissue to insulin followed by washing results in a considerable binding of the hormone by the tissue. Fractional centrifugation of such tissues that have bound hormone, shows that the hormone is located in the cell membrane fraction. We have mentioned that the disulphide bridge of the molecule is necessary for biological activity and it has been suggested that insulin might be 'used up' by interaction of the hormone with the sulphydryl groups of the cysteine radicals of the proteins of cell membranes.

Peptide	HCL secretion		Gastric motility		Intestinal motility		Pancreatic secretion		Pepsin secretion
	stim.	inhib.	stim.	inhib.	stim.	inhib.	vol.	enzyme	stim.
Glu. Gly. Pro. Try. Met. (Glu)5. Ala. Tyr. Gly. Try. Met. Asp. Phe–NH2	+	+	+	O	+	+	+	+	+
Z. Asp. Phe–NH2	O	O	O	O	O	O	O	O	O
Z. Met. Asp. Phe–NH2	O	O	O	+	O	O	O	O	O
Z. Try. Met. Asp. Phe–NH2	+	+	+	+	+	+	+	+	+
Z. Gly. Try. Met. Asp. Phe–NH2	+	+	+	?+	+	+	+	+	+
Z. Gly. Try. Met. Asp. Phe–OMe	O	O	O	O	O	O	+	+	?+

Z = Benzyloxicarbinoyl   + = stimulates   O = no effect   ? = doubtful or small effect

FIGURE 51. Physiological effects of some peptides related to gastrin.
Data from Gregory, R. A. & Tracy, Hilda, J. (1966). Studies on the chemistry of gastrins 1 and 2. Gastrin ed. M. I. Grossmen, Butterworth, London.

Structure–activity relationships.—4. Gastrin

During the work of synthesis of the polypeptide hormone gastrin (see p. 167) a great number of different molecules had been made. In order to find out how many of the seventeen constituent amino acids of the gastrin molecule were necessary to produce the distinctive physiological effects on the alimentary tract a range of synthetic molecules was prepared. It was found that only four residues are really essential to mimic the physiological actions of the complete molecule. These four are tryptamine–methionine–aspartic acid–phenylalanine NH_2 (fig. 51). Although the entire range of physiological activities displayed by natural gastrin is possessed by this tetrapeptide sequence the potency of the tetrapeptide is considerably less than that of natural gastrin. Thus the power to stimulate gastric acid secretion by the tetrapeptide is only about one-fifth of that of the whole molecule of gastrin.

As the peptide chain is progressively lengthened, the general potency of the molecule increases. The absence of the terminal NH_2 group radically alters the potency of the hormone and this has led to speculation that inactivation of the hormone in the body may be brought about by removing this amide.

Structure–activity relationships.—5. Acetylcholine

Acetylcholine is a local hormone which acts as a neurotransmitter substance in various tissues. The hormone is released from all somatic motor-nerve terminals, at all preganglionic terminals in the sympathetic and parasympathetic nervous system, at all postganglionic terminals in the parasympathetic nervous system and at some postganglionic terminals in the sympathetic nervous system. Its release is associated with changes in the ion permeability of the cell membrane of the target cells which results in electrical changes in the cells. Although it produces a common effect in the various tissues the nature of the receptive substance and its union with the hormone is probably of a different nature in different classes of tissues. This view is based upon several kinds of evidence. A variety of chemical compounds, both naturally occurring and synthetic, can mimic the effect of acetylcholine. This mimicry is, however, incomplete. The alkaloid muscarine simulates the action of acetylcholine only on smooth and cardiac muscle and exocrine glands. The actions of acetylcholine on these structures is thus called 'muscarinic' to distinguish them from other effects. Similarly the alkaloid nicotine has a selective action in mimicking the effects of acetylcholine on autonomic ganglia and skeletal muscle on which it produces an initial stimulation followed later by paralysis—an effect shown also by acetylcholine in high doses. The actions of acetylcholine on autonomic ganglia and skeletal muscle is described as 'nicotinic'. This distinction between two classes of acetylcholine–receptor interaction is also supported by the fact that drugs which can block the action of acetylcholine on the target organs also fall into two classes. One class of compounds exemplified

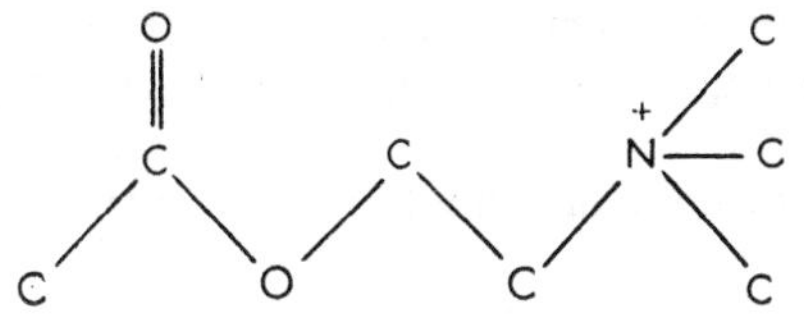

Acetylcholine

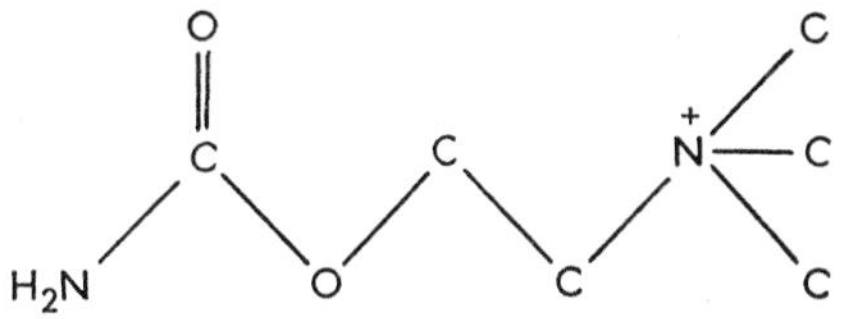

Carbaminoylcholine

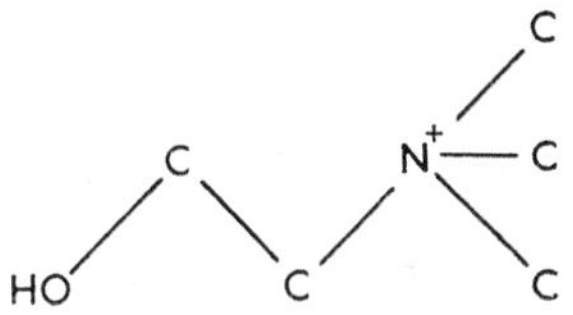

Choline

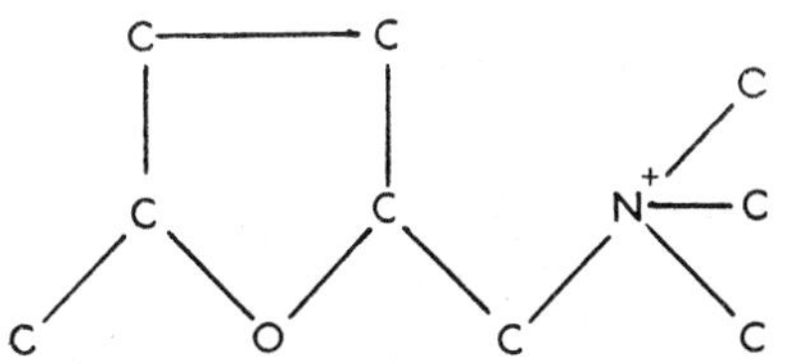

Methylfurthrethonium

FIGURE 52. Some substances mimicking the action of acetylcholine on the bowel. Note the wide variation in structure of the chain of the molecule. All the substances possess the positively charged terminal onium group.

by the alkaloid atropine can block the muscarinic effect of acetyl-choline, while another class, notably curare and 'curariform' drugs, block the action on skeletal muscle.

The onium group in acetylcholine

The positively charged onium group of acetylcholine appears to be important in determining the excitatory effects of the hormone. A large number of quaternary ammonium compounds bearing the positively charged onium group can mimic the effects of acetyl-choline. Fig. 52 shows some of these compounds which mimic the nicotinic actions of the hormone on smooth muscle of the bowel. It

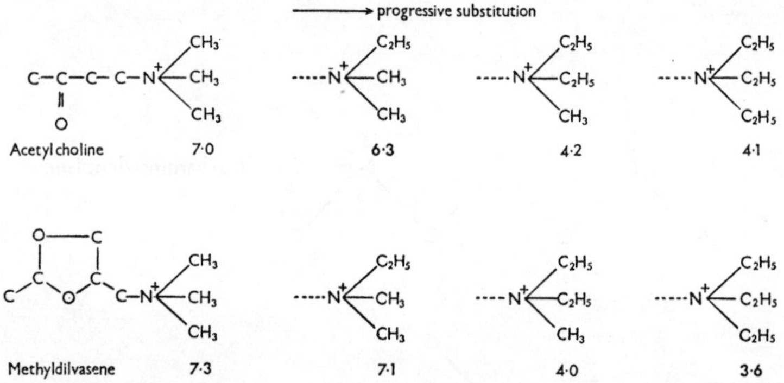

FIGURE 53. The effect of ethyl substitution of the methyl groups in acetylcholine and an analogue, methyldilvasene, on the affinity of the hormone for the receptor. Quantitative estimates of the affinity are shown which reflect the tendency of the hormone and receptor to unite. The effect of a reduction in affinity is shown by the increasing concentration of the activator which is necessary to produce the same effect in a target tissue.

Data from Ariëns, E. J. et al. (1964). Drug-receptor interaction: interaction of one or more drugs with one receptor system. In *Molecular Pharmacology*, ed. Ariëns, E. J., Academic Press.

is expected that the receptor substance for acetylcholine in the target cells will show a complementary structure to the hormone, and in respect of the positively charge onium group a negatively charged site on the receptor is anticipated. Because the force generated by inter-action of heavily charged ionic groups (acetylcholine and receptor) have a wide range of influence the cationic head of the acetylcholine molecule (the onium group) may be important in 'seeking out' tissue receptors and guiding the hormone on to the receptor. There are other forces of interaction, e.g. van der Waals' forces and hydrogen bond formation, but these have a shorter range of influence. They may well be important in binding the hormone after the cationic head has approached the negative site on the receptor.

The importance of the cationic head of the molecule has been shown in various ways. A reduction in the positive charge on the onium group, shielding of the electrical fields around the onium group by appropriate chemical substitution in the molecule or 'steric hindrance' produced by substitution of large chemical groups on the onium group all reduce the potency of the hormone. Thus substitution of the methyl groups in the onium group by the larger ethyl group reduces the activity of acetylcholine. This effect of ethyl substitution may be due to various factors. There may be masking of the positive charge on the onium group or interference by the electrical fields around the ethyl group with the negative site on the receptor. The effect of ethyl substitution is progressive and increases with the number of substituted groups (fig. 53).

The chain in acetylcholine

In addition to the onium group the chain of the molecule may be important in binding the hormone to the receptor by means of those short range forces already mentioned. Before these forces can come into operation the hormone must come into close relationship with the receptor. It is to be expected that changes in the spatial structure of the molecule which hinders this close relationship will result in a reduction in its biological activity. When this occurs much higher doses of the hormone are needed to produce the same physiological effect.

The requirements of the structure of the chain for optimal biological activity have been found to vary from one target tissue to another, supporting the view that there must be differences in the structure of the receptor in these tissues. The action of acetylcholine at vagal nerve terminals requires a chain of five atoms for optimal effects and a reduction in length to four atoms results in a marked loss in activity. However, at the ganglionic synapse a side chain of four atoms is optimal.

Summary

In the above examples we have tried to introduce the concept that various parts of the hormone molecule may perform different but complementary functions. Examples have been given of the kind of investigations which can be used in this study of the 'functional topography' of the hormone molecule.

Bibliography

Techniques in Endocrinology
* Dorfman, R. I. (1962) ed. *Methods in Hormone Research*, Vols. 1 and 2. Academic Press, New York.
* Eckstein, P. & Knowles, F. (1963) eds. *Techniques of Endocrine Research*. Academic Press.

Progesterone as a local hormone
* Csapo, A. (1961). Defence mechanism of pregnancy. *Ciba Foundation Study Group*, No. 9, 3–27. Churchill, London.
  Goto, M. & Csapo, A. (1959). The effect of the ovarian steroids on the membrane potential of uterine muscle. *J. gen. Physiol.* 43, 455–466.
* Schofield, Brenda M. (1963). The physiology of the myometrium. *Recent Advances in Physiology*, 222–251. Creese, R., ed. Churchill, London.
  Schofield, Brenda M. (1966). The local influence of the placenta on myometrial activity in Endogenous substances affecting the myometrium, *Memoirs of the Society for Endocrinology*, 14. Pickles, V. R. & Fitzpatrick, R. J., eds. Cambridge University Press.

Mechanisms of hormone action
* Clever, U. (1964). Actinomycin and Puromycin: effects on sequential gene activation by ecdysone. *Science.* 146, 794–795.
* Karlson, P. (1965) ed. *Mechanisms of hormone action.* Academic Press, New York.
* Kroeger, H. & Lezzi, M. (1966). Regulation of gene action in insect development. *A. Rev. Entomol.* 11, 1–22.
  Levine, R. (1964). Analysis of the glucose transport theory of insulin action. *Proc. 2nd. Int. Congr. Endocrinol.* 30–34. Excerpta Medica, London.
* Mongar & de Reuck (1962) eds. *Ciba Foundation Symposium on Enzymes and Drug Action.* Churchill, London.
  Sekaris, C. E. & Lang, N. (1964). Induction of Dopa-decarboxylase activity by insect messenger RNA in an in vitro amino-acid incorporating system from rat liver. *Life Sciences.* 3, 625–632.
* Spickett, S. G. (1967) ed. Memoirs of the Society for Endocrinology, No. 15. *Endocrine Genetics.* Cambridge University Press.
  Sutherland, E. W. & Robinson, G. A. (1966). The role of cyclic-3′, 5′-AMP in responses to catecholamines and other hormones. *Pharmacol. Rev.* 18, 145–159.

* Represents textbooks or review articles.

208

Hormones and calcium metabolism

Copp, D. H., Cameron, E. C., Cheney, B. A., Davidson, A. G. F. & Henze, K. G. (1962). Evidence for calcitonin—a new hormone from the parathyroid that lowers blood calcium. *Endocrinology.* **70**, 638–649.

Foster, G. V., Baghdiantz, A., Kumar, M. A., Slack, E., Soliman, H. A. & Macintyre, I. (1964). Thyroid origin of calcitonin. *Nature, Lond.* **202**, 1303–1305.

* Geschwind, I. I. (1961). Hormonal control of calcium, phosphorus, iodine, iron, sulphur, and magnesium metabolism. In Comar, C. L. & Bronner, F. eds. *Mineral Metabolism 1–B.* Academic Press, New York.

* Greep, R. O. & Talmage, R. V. (1961) eds. *The Parathyroids.* Charles C. Thomas, Springfield, Ill.

Macintyre, I., Parsons, J. A. & Robinson, C. J. (1967). The effect of thyrocalcitonin on the blood-bone calcium equilibrium in the perfused tibia of the cat. *J. Physiol.* **191**, 393–405.

Hormones and sodium and water metabolism

Clark, Barbara J. & Rocha e Silva, M. (Jr.) (1967). An afferent pathway for the selective release of vasopressin in response to carotid occlusion and haemorrhage in the cat. *J. Physiol.* **191**, 529–542.

* Davis, J. O. (1962). The control of aldosterone secretion. *Physiologist.* **5**, 65–86.

* Gauer, O. H. & Henry, J. P. (1963). Circulatory basis of fluid volume control. *Physiol. Rev.* **43**, 423–481.

Ginsburg, M. & Brown, L. M. (1957). The effects of haemorrhage and plasma hypertonicity on the neurohypophysis.* In *The Neurohypophysis*, Heller, H. ed. Butterworths, London.

Henry, J. P. & Pearce, J. W. (1956). The possible role of atrial stretch receptors in the induction of changes in urine flow. *J. Physiol.* **131**, 572–585.

* O'Connor, W. J. (1962). Renal Function. *Memoirs of the Physiological Society*, 11. Arnold, London.

* Peart, W. S. (1965). The renin-angiotensin system. *Pharmacol. Rev.* **17**, 143–182.

* Pickford, Mary (1945). Control of the secretion of the antidiuretic hormone from the pars nervosa of the pituitary gland. *Physiol. Rev.* **25**, 573–595.

Share, L. (1965). Effects of carotid occlusion and left atrial distention on plasma vasopressin. *Am. J. Physiol.* **208**, 219–223.

* Williams, P. C. (1963) ed. Hormones and the Kidney. *Mem. Soc. Endocrinol.* No. 13. Academic Press, London.

Chromaffine tissue, the adrenal medulla, sympathetic nervous system

Coupland, R. E. (1965). The chromaffine cell. *Science.* **10**, 52–58.

* Malmejac, R. J. (1964). Activity of the adrenal medulla and its regulation. *Physiol. Rev.* **44**, 186–218.

* Sutherland, E. W. & Robison, G. A. (1966). The role of cyclic-3',
    5'-A.M.P. in responses to catecholamines and other hormones.
    Pharmacol. Rev. **18**, 145–159.
* Vane, J. R., Wolstenholme, G. E. W. & O'Connor, M. (1960) eds.
    Adrenergic Mechanisms. *Ciba Foundation Symposium.* Churchill,
    London.

The adrenal cortex

* Dorfman, R. I. & Ungar, F. (1965). *Metabolism of Steroid Hormones.*
    Academic Press, New York.
  Hamburg, D. A. & Kessler, S. (1967). A behavioural–endocrine–
    genetic approach to stress problems. In *Endocrine Genetics.* Spickett,
    S. G., ed. Cambridge University Press.
* Lardy, H. A. (1966). Gluconeogenesis: Pathways and Hormonal Regu-
    lation. The Harvey Lecture, Series 60, 261–278.
* Von Euler, U. S. & Heller, H. (1963) eds. *The Adrenocortical Hormones.*
    Academic Press, New York.
  Weber, G. C., Srivastava, S. K. & Singhal, R. L. (1965). Role of hor-
    mones in homeostasis, VII. Early effects of corticosteroid hormones on
    hepatic gluconeogenesis, ribonucleic acid metabolism and amino acid
    level. *J. Biol. Chem.* **240**, 750–756.
* Wolstenholme, G. E. W. & Porter, Ruth (1967) eds. The Human
    Adrenal Cortex: Its function throughout life. *Ciba Foundation Study
    Group*, No. 27. Churchill, London.

Hormones and temperature regulation

  Anderson, B., Ekman, L., Gale, C. D. & Sundsten, J. W. (1963). Control
    of thyrotrophic hormone 'TSH' secretion by the 'heat loss centre'.
    Acta. physiol. scand. **59**, 12–23.
* Benzinger, T. H. (1964). The thermal homeostasis of man. In *Homeo-
    stasis and feedback mechanisms.* Hughes, G. H. ed. Cambridge Uni-
    versity Press.
  Brittain, R. T. & Handley, S. L. (1967). Temperature changes produced
    by the injection of catecholamines and 5-hydroxytryptamine into the
    cerebral ventricles of the conscious mouse. *J. Physiol.* **192**, 805–813.
* Carlson, L. D. (1966). The role of catecholamines in cold adaptation.
    Pharmacol. Rev. **18**, 291–301.
  Feldberg, W. & Myers, R. D. (1964). Effects on temperature of amines
    injected into the cerebral ventricles. A new concept of temperature
    regulation. *J. Physiol.* **173**, 226–237.
  Hammel, H. T., Jackson, D. C., Stolwisk, J. A. J., Hardy, J. D. &
    Stromme, S. B. (1963). Temperature regulation by hypothalamic
    proportional control with an adjustable set point. *J. appl. Physiol.* **18**,
    1146–1154.
* Hardy, J. D. (1961). Physiology of temperature regulation. *Physiol. Rev.*
    41, 521–606.

* Hart, J. S. (1964). Insulative and metabolic adaptations to cold in vertebrates. *Symposium, Society for Experimental Biology.* **18**, 31–48.
* Pitt Rivers, R. & Tata, J. R. (1959). *The thyroid hormones.* Pergammon Press, New York.
* Von Euler, C. (1961). Physiology and pharmacology of temperature regulation. *Pharmacol. Rev.* **13**, 361–398.

Hypothalamus and pituitary gland

Averill, R. L. W., Solomon, D. F. & Worthington, W. C. (Jr.) (1966). Thyrotrophin releasing factor in hypophysial-portal blood. *Nature, Lond.* **211**, 144–145.
* Burns, R. N. (1961). Role of hormones in the differentiation of sex. In *Sex and Internal Secretion*, 76–158. W. C. Young, ed. The Williams and Wilkins Company, Baltimore.
Green, J. D. & Harris, G. W. (1949). Observations of the hypophysial-portal vessels of the living rat. *J. Physiol.* **108**, 359–361.
* Guillemin, R. (1964). Hypothalamic factors releasing pituitary hormones. *Recent Prog. Horm. Res.* **20**, 89–130.
* Harris, G. W. (1955). *Neural Control of the Pituitary Gland.* Edward Arnold, London.
* Harris, G. W. (1962). The development of neuroendocrinology. In *Frontiers in Brain Research*, pp. 191–241. French, J. D., ed. Columbia University Press, New York and London.
Harris, G. W. & Levine, S. (1965). Sexual differentiation of the brain and its experimental control. *J. Physiol.* **181**, 379–400.
Harris, G. W. & Michael, R. P. (1964). The activation of sexual behaviour by hypothalamic implants of oestrogen. *J. Physiol.* **171**, 275–301.
Palka, Y. S. & Sawyer, C. H. (1966). The effects of hypothalamic implants of ovarian steroids on oestrus behaviour in rabbits. *J. Physiol.* **185**, 251–369.
* Sawyer, C. H. (1964). Control of secretion of gonadotrophins. In *Gonadotrophins*, 113–159. Cole, H. H., ed. W. H. Freeman and Co., San Francisco.
Worthington, W. C. (Jr.) (1966). Blood samples from the pituitary stalk of the rat: method of collection and factors determining volume. *Nature, Lond.* **210**, 710–712.

Hormones and gastrointestinal activity

* Fenton, P. F. (1960). The stomach and small intestine. In *Medical Physiology and Biophysics*, 924–943. Ruch, T. C. & Fulton, J. F., eds. 18th edition. W. B. Saunders Company.
* Gregory, R. A. (1962). Secretory mechanisms of the gastro-intestinal tract. *Monographs of the Physiological Society*, no. 11. Edward Arnold.
* Grossman, M. I. (1966) ed. Gastrin-proceedings of a conference. Butterworths.

Hormones and reproduction

Diamond, L. K., Jacobson, W. & Sidman, R. L. (1967). Short communication: testosterone-induced D.N.A. synthesis in human bone marrow. *Memoirs of the Society for Endocrinology, No. 15, Endocrine Genetics.* Spickett, S. G., ed. Cambridge University Press.

Liao, S. (1965). Influence of testosterone on template activity of prostate ribo-nucleic acids. *J. biol. Chem.* **240**, 1236–1243.

* Nalbandov, A. V. (1958). *Reproductive Physiology.* Freeman, San Francisco, California.

* Villee (1961) ed. Control of ovulation. Pergamon Press, London.

Wicks, W. D. & Kenny, F. T. (1964). R.N.A. synthesis in rat seminal vesicles: stimulation by testosterone. *Science.* **144**, 1346–1347.

Wilson, J. D. & Loeb, P. M. (1965). Localization of testosterone-1, 2-^{3}H in the preen gland of the duck. *J. clin. Invest.* **44**, 1111.

* Wolstenholme, G. E. W. & O'Connor, M. (1967) eds. Endocrinology of the Testis. Churchill, London.

* Young, W. C. (1961) ed. Sex and Internal Secretions. The Williams and Wilkins Company, Baltimore.

* Zuckerman, S. (1962) ed. *The Ovary.* Vols. 1 and 2. Academic Press, London.

See also references for progesterone as a local hormone and for hypothalamus and pituitary gland.

The chemical structure of hormones in relation to their biological activity

* Ariëns, E. J. (1964) ed. Molecular Pharmacology: The Mode of action of biologically active compounds. (2 Vols.) *Medicinal Chemistry,* Vol. 3. de Stevens, G., ed. Academic Press.

Sanger, F. (1960). Chemistry of insulin. *Brit. med. Bull.* **16**, 183–188.

* Spickett, S. G. (1967) ed. *Memoirs of the Society for Endocrinology, 15, Endocrine Genetics.* Cambridge University Press.

Index

213